THE BETHLEHEM OIL MILL 1745-1934

German Technology in Early Pennsylvania

1. This scale model shows the interior of the Bethlehem oil mill as it looked in 1766. The left water-wheel (center) powered the oilseed mill (lower left) and the groat mill (upper left). The right waterwheel drove the hemp, tanbark, and snuff stamping mills (lower right). This model was constructed by Stephen Young in 1975 for display at the restored Moravian tannery in Bethlehem.[1]

THE BETHLEHEM OIL MILL 1745-1934

OILSEED MILL

HEMP MILL

TANBARK MILL

GROAT MILL

SNUFF MILL

WATERWORKS

German Technology in Early Pennsylvania

CARTER LITCHFIELD

HANS-JOACHIM FINKE

STEPHEN G. YOUNG

KAREN ZERBE HUETTER

OLEARIUS EDITIONS

Published and distributed by

OLEARIUS EDITIONS
Drawer H
Kemblesville
Pennsylvania 19347
U.S.A.

ISBN 0-917526-02-3
Library of Congress Card No. 82-61069

Printed and bound in the United States of America

CONTENTS

PREFACE

Our understanding of mill technology in eighteenth century America has always been hindered by a shortage of original sources. Colonial millwrights and machinists often built without formal plans, and they rarely prepared written descriptions of their accomplishments. The machinery that operated in eighteenth century mills has seldom survived in situ. Hence today's interpretation of American colonial mill technology relies to a large extent upon secondary descriptions, contemporary European practices, and the analysis of fragmentary artifacts.

The Moravian religious and industrial community at Bethlehem, Pennsylvania is a fortunate exception. From its founding in 1741 as a missionary colony of the *Unitas Fratrum* (Moravian) Church in Herrnhut, Germany, the Moravians kept extensive written records of all their activities. In addition they regularly sent detailed reports on their work back to their European sponsors. Manufacturing was an important part of the Bethlehem economy, so the construction and operation of their mills were recorded in great detail. Fortunately these records have survived almost intact at the Moravian Archives in Bethlehem and at the *Archiv der Brüder-Unität* in Herrnhut. American technological historians have made only limited use of these documents, however, because most of them are written in the unfamiliar German gothic script.

From this large store of information we have selected the Bethlehem oil mill (1745-1934) for a detailed historical and technological study. Although named for its principal operation, pressing linseed oil, actually five different processes were carried out there. The building contained an oilseed mill, two hemp mills, a tanbark mill, a groat mill, and a snuff mill.

As a basis for our investigation the entire diary and the account book kept during the construction of the 1765 oil mill building have been translated into English. The 1766 architectural and machinery drawings of the building have been located. Numerous inventories and ledgers from the oil mill industries have been studied in detail.

Extensive newspaper accounts describing the 1832-1913 operation of the Bethlehem waterworks at the building have been thoroughly reviewed. Many other sources of relevant information have been searched.

From these varied sources we have synthesized the biography of this early American industrial building and reconstructed the manufacturing processes used there. We also present our translation of the original construction records for the 1765 building so that the reader can view the mill through the words of the men who built it. Together, these two viewpoints provide a unique insight into how an early American industrial mill was built and operated.

INDUSTRY IN EARLY BETHLEHEM

MORAVIAN COLONISTS

Bethlehem, Pennsylvania, the major settlement of the Moravians in North America, was founded in 1741. The Moravians, an early Protestant group originating in central Europe, came to America to preach the gospel to the other German settlers, to the Indians, and to Blacks.[1] Located on the Lehigh River 47 miles (75 km) north of Philadelphia, Bethlehem served as their headquarters for missionary activities throughout the colonies.[2]

The Moravians trace their origins back to Jan Hus, a Catholic priest of Prague, who was burned at the stake for heresy in 1415. Some of Hus' followers who lived in Bohemia and Moravia (today Czechoslovakia) formed their own church, the *Unitas Fratrum* (Unity of the Brethren), in 1457. After worshipping openly for many years, they were forced underground after the Thirty Years' War (1618-1648).

In 1722 Count Nicholas von Zinzendorf, a devout 22-year-old Lutheran nobleman, allowed some of them to settle on his estate at Berthelsdorf near Dresden, Germany. There the Brethren, joined by others from all over Europe, built a new religious community called Herrnhut that emphasized the pietistic experience.

Count Zinzendorf, the group's patron, had important contacts in other countries; and through his influence a strong commitment to missionary service was established. As the movement grew, many new Moravian communities were founded; and Herrnhut became the center for their activities throughout the world. These new communities maintained strong ties with Herrnhut, looking to the mother church there for spiritual and secular guidance, personnel, and financial aid.[3]

The Moravians who founded Bethlehem in 1741 were practical as well as pious; many were experienced in specific crafts and trades. At Bethlehem they sought to establish a permanent, self-sufficient, religious community that could support their planned missionary activities.

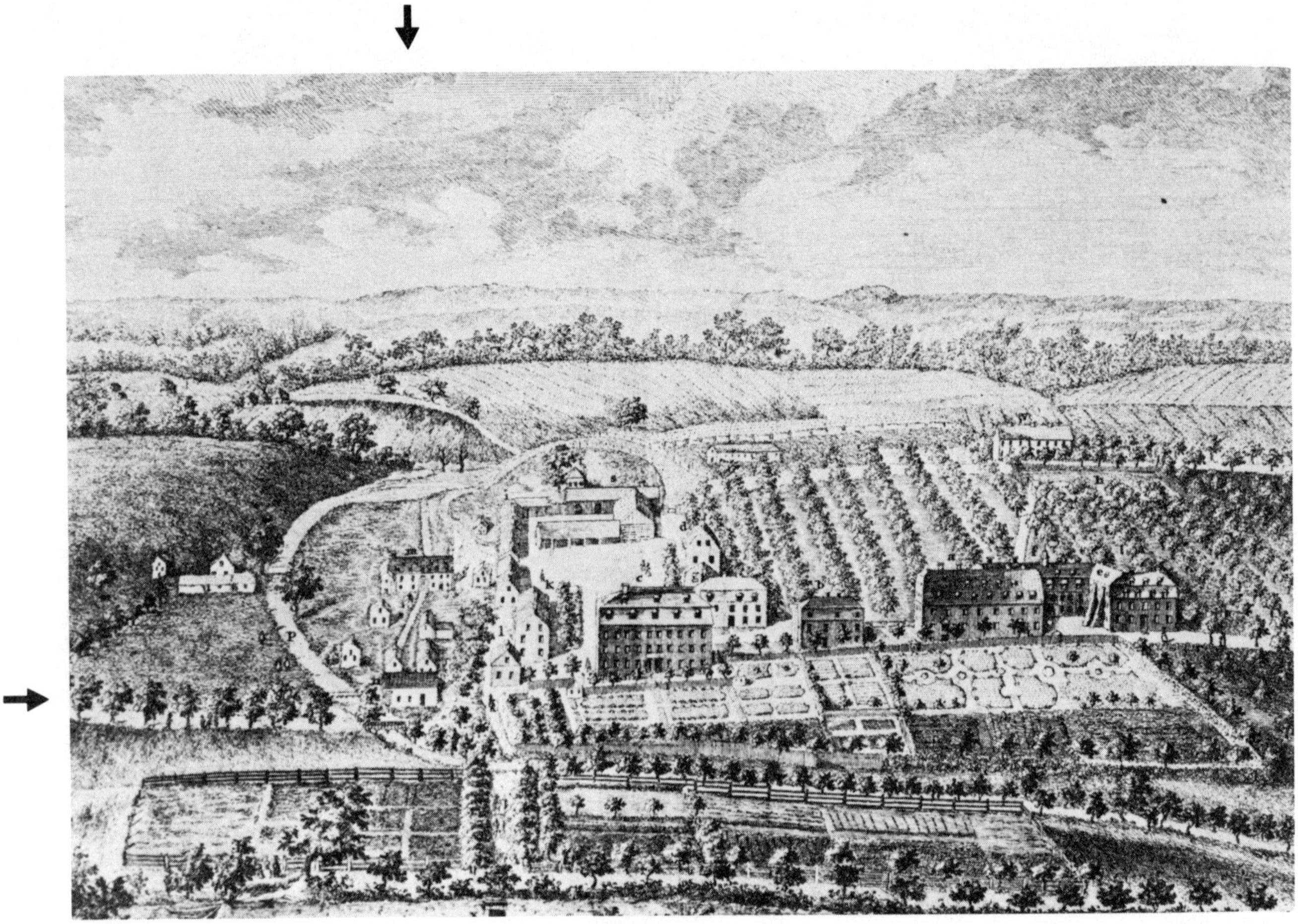

2. *Nicholas Garrison Jr.'s drawing of Bethlehem in 1757 shows a small, ordered community surrounded by orchards, gardens, and fields. The industrial area along Monocacy Creek lies at left center with the 1752 oil mill indicated by arrows.*[6]

They were skilled builders, who soon created an impressive community on the colonial frontier. Israel Acrelius, a Swedish minister who visited Bethlehem in June 1754, reported:

> As one comes on the road from Philadelphia . . . and . . . finally draws nigh to Bethlehem, reaching a hill two miles away from it, there it shows itself in a glory which is not much inferior to that of Konungahof [a summer residence of the Swedish king].[4]

One of the earliest published views of Bethlehem (figure 2), drawn in 1757 by Nicholas Garrison Jr.,[5] shows a small, ordered community in a neat, rural setting. The residential section with its massive stone buildings stood along the crest of the hill overlooking the Lehigh River. This row of buildings formed the major east-west axis of the community (the present Church Street). Each building housed a specific "choir," a communal living group whose membership was determined by age, sex, and marital status. Many small workshops were located in the Single Sisters' (1744) and the Single Brethren's (1748) choir houses, while other manufacturing establishments were situated along Monocacy Creek to the west. To the north were various barns and stables. Meadows, grain fields, extensive fruit orchards, and herb, vegetable, and flower gardens surrounded the town.

The religious impetus for Bethlehem's founding had a profound influence on the economy of the new community. The leaders of early Bethlehem strove for economic self-sufficiency, believing that they could better support their missionary activities and also be less affected by outside secular influences if they provided most of their own material needs. In June 1742 the population was formally divided into two groups.[7] Members of the Pilgrim Congregation were assigned to the mission field while residents in the Home Congregation remained in Bethlehem to support the missionary work. There was a continual interchange between the two groups, with the Pilgrims always constituting a sizable portion of the community. In 1751, for example, the Moravian population consisted of 744 individuals. Eighty-eight of them, approximately one-sixth of the adults, preached the gospel in places away from Bethlehem.[8]

Carefully regulated economic activity was, therefore, essential to assure the prosperity of Bethlehem as well as to support the numerous missionaries in other locations. During the first twenty years all residents worked under a communal system called the General Economy. The church owned all the land, buildings, furnishings, livestock, trades, tools, and the goods produced. Rather than working for individual gain, all members worked for the community, receiving food, clothing, shelter, education, and cultural benefits in return. No task was considered too menial or unimportant, because it helped the community reach its common goal of winning souls for Christ.

By 1762 the growing size and complexity of Bethlehem brought about a modification of the General Economy. A new system permitting both individual and congregational enterprises was instituted. The Moravian Church continued to own all the land and many of the buildings, but craftsmen now rented their workshops and sold their wares for a profit. The *Aufseher Collegium*, a committee of respected community leaders, oversaw the daily operation of the trades and other secular affairs. In addition, Bethlehem's married residents began to live as families in their own dwellings. The single women, single men, and widows continued to reside in their respective choir houses, but they now paid room and board. Parents also paid tuition to send their children (both boys and girls) to excellent schools that had been established in Bethlehem and in the nearby Moravian settlement of Nazareth.[9]

Most early visitors to Bethlehem were surprised to find such a variety and quality of goods produced in a small frontier settlement. As early as 1747 some forty crafts had been established in the town.[10] When Thomas Pownall, a prominent colonial administrator, visited Bethlehem in 1754, he reported the following trades:

> There were, when I was there, the Following Trades carried on by the Fratres at this Settlement. Saddle-tree maker, Sadler, Glover, Shoemaker,

> Stocking-weavers, 4 frames going, Button maker, Taylor & Women Taylor, Hatter, Ribband-weavers, Linnen-weavers, 6 looms in work, Woollen-weavers, three looms at work, Wool-comber, Dyer, Fuller, Dresser, Tanner, Currier, Skinner, Butcher, Miller, Chandler, Oil-maker, Baker, Cooper, Joiner, Carpenter, Mason, Glazier, Brick maker, Stone Cutter, Turner, Potter, Stovemaker, Wheelwright, Blacksmith, Gunsmith, Nail-Maker, Lock-smith, Pewterer, Tinman, Silver-smith, Clockmaker, Harnes-maker, Hemp dresser, Boat-builder, Surgeon, Apothecary. These Artificiers & Manufacturers had each separate appartments & shops to themselves.[11]

By the early 1750's, the Moravians were producing enough cloth to supply the clothing needs of their members,[12] an important goal for any self-sufficient eighteenth century community. When the first general store opened on Market Street in 1753, the Brethren were able to produce or raise 180 of the 200 items offered.[13] While many of these products were consumed by the Moravian community, some were sold to neighbors while others, such as linseed oil, were marketed in Philadelphia and New York.[14] The Moravians used the resulting cash to purchase materials like gunpowder and salt, which they did not produce, or to pay their taxes to the colonial Pennsylvania government.

The emphasis placed on the development of handicraft technology rather than agricultural pursuits was an integral part of their plan for extensive missionary activities. Moravian leaders believed that men and women engaged in trades had more mobility than their counterparts working in agriculture. Farmers, by the seasonal nature of their occupation, remained tied to the soil. Since freedom of movement was essential for missionary work, craftsmen were more suited to this calling.[15]

MONOCACY INDUSTRIAL AREA

The Moravians who came to Bethlehem brought with them the technological heritage of Europe and utilized it to great advantage. They built many mills and manufacturing establishments along Monocacy Creek on the west side of town where ample water and waterpower were available.[16] A 1772 map of this industrial area (figure 3) shows a soapworks, dyehouses, a combined grist and fulling mill,[17] a butchery, a tannery, a tawery, a spring house, a tanbark peeling shed, the waterworks, the oil mill, an oat drying kiln, a smithy, nailsmith and locksmith shops, a pottery, and a cabinet shop. Many of these Monocacy Creek trades were originally housed in small log structures. As Bethlehem grew and its needs expanded, however, most of the log buildings were replaced with larger, permanent limestone structures in the traditional Germanic style of architecture.[19]

Tanning was one of the most profitable trades for the Moravians. Their tannery supplied the needs of shoe-

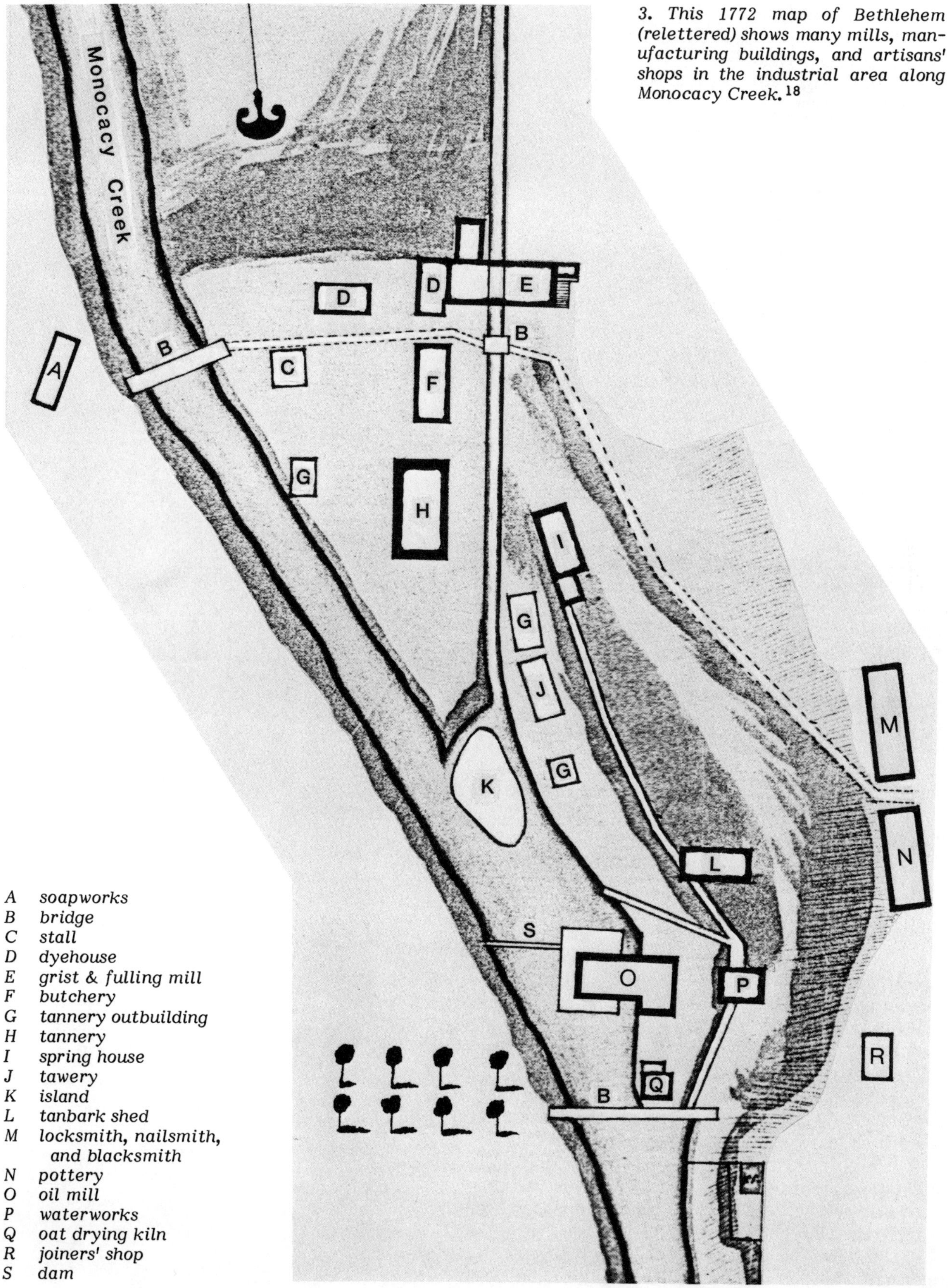

3. *This 1772 map of Bethlehem (relettered) shows many mills, manufacturing buildings, and artisans' shops in the industrial area along Monocacy Creek.*[18]

A soapworks
B bridge
C stall
D dyehouse
E grist & fulling mill
F butchery
G tannery outbuilding
H tannery
I spring house
J tawery
K island
L tanbark shed
M locksmith, nailsmith, and blacksmith
N pottery
O oil mill
P waterworks
Q oat drying kiln
R joiners' shop
S dam

makers, harnessmakers, and saddlers not only in Bethlehem and Nazareth but in the surrounding areas as well. The first tannery was built in 1743, and in 1761 a large limestone building still extant was constructed to house the expanding industry.[20] By the 1760's the Moravian tanners were working 1000-2000 hides annually.[21] During the American Revolution the annual output rose sharply.[22] The war ended trade with England, the major supplier of leather to the colonies, and forced local tanners to increase production. The needs of the Continental Army for goods like leather boots, caps, coats, buckets, and saddles also created a demand for greater output.[23]

The first water-powered trade along Monocacy Creek was a grist mill built in 1743 (less than two years after Bethlehem was founded) to meet the Moravians' basic need for flour.[17] The grist mill was rebuilt in 1751; it originally contained one run of grinding stones, but increased usage by both Bethlehem residents and neighboring settlers led to a second run being added in 1753. The new building also included a fulling mill for processing woolen fabrics. Four experienced Moravian weavers had arrived from England in 1749 and had started the manufacture of woolen cloth in Bethlehem.[24]

The main source of drinking water for the town was a copious spring located near Monocacy Creek in the industrial area. The problems involved with transporting this water supply to the residential area located at the top of the steep hill above the creek provided the impetus to build a waterworks powered by a waterwheel. Hans Christensen, the millwright who figured prominently in numerous Moravian mill projects, experimented with a pumping system in 1754 but did not complete the work because the pipes kept leaking. However, in 1761-62 he constructed a new 3-cylinder pumping system and a new waterworks building. This new pump successfully forced spring water up the hill to a wooden collecting tower located about 94 feet (29 m) above the level of the pump. From there gravity carried the water to cisterns throughout the town.[25] This European technology for municipal water distribution was not matched in America until the Philadelphia water system was completed in 1801.[26]

The most complex mill built in the Monocacy industrial area was the 1765 oil mill. Although named for its principal operation, oilseed pressing, other processes were also located in this building so that they could utilize the same waterpower system. Machinery for pressing linseed oil from flaxseed, stamping tanbark, softening hemp, grinding groats, and milling snuff were operated there. The history and technology of this remarkable colonial manufacturing complex are explored in this book.

Eighteenth century visitors were usually shown the industrial area along Monocacy Creek with great pride, and it obviously proved to be an impressive sight. When future President John Adams visited Bethlehem in April 1777, he wrote to his wife Abigail about the Moravians saying:

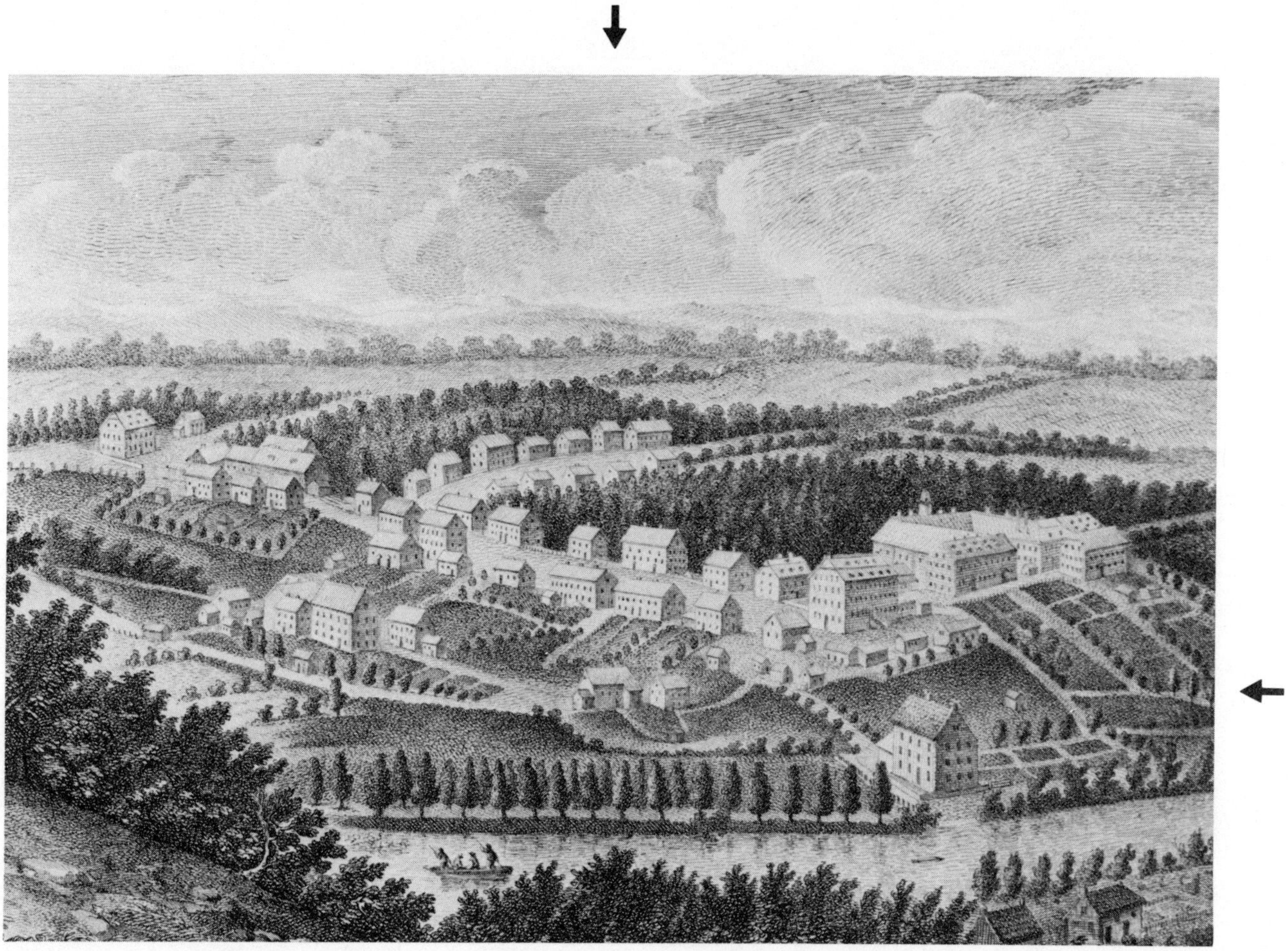

4. In 1784 Garrison and Oerter drew this view of Bethlehem from the southwest.[28] Comparison with Garrison's 1757 drawing (figure 2) shows how the community had expanded to the north and east. The 1765 oil mill stands at the center foreground as indicated by arrows.

They have carried the mechanical Arts to greater Perfection here than in any Place which I have seen. . . . They have a fine Sett of Mills. The best Grist Mills and bolting Mills, that are any where to be found. The best fulling Mills, an oil Mill, a Mill to grind Bark for the Tanyard, a Dying House where all Colours are died, Machines for shearing Cloth &c.[27]

TWO CENTURIES OF CHANGE

During the first half of the nineteenth century outside influences, including construction of the Lehigh Canal to transport coal from northeastern Pennsylvania and the discovery of zinc deposits south of the Lehigh River, brought about many changes in the economic and social structure of the town. In the 1840's the Moravian Church, which still owned all the land and some of the buildings in Bethlehem, found itself in great financial difficulty and decided to sell much of its property. Thus in 1844 the town was opened to non-Moravian residents and workers for the first time.

The Monocacy industrial area remained economically important only until the 1840's. Most Moravian tradesmen apparently lacked either the capital or the

5. The large D. &. A. Luckenbach Bethlehem Roller Flour Mills dominated the industrial area along Monocacy Creek at the end of the nineteenth century.[29] *This fine Victorian building was built in 1869 to replace the old 1751 mill which had burned.*

6. By the end of the nineteenth century, the old Moravian tannery was no longer in operation but had been converted to residences with various sheds and outbuildings.[30]

incentive to expand their businesses or to introduce mass production. The dearth of local tanbark and the high price of shipping it in from western Pennsylvania helped force the tannery out of business. It finally closed in 1873 after limited use for the previous thirty years. In 1832 the water pumping operation was switched from the old 1762 waterworks into the larger oil mill building standing a few feet to the west. The oil mill served as Bethlehem's pumping station until 1913 when the nearby spring from which it obtained water was declared contaminated.

Only the grist mill continued as a profitable enterprise. When the 1751 mill burned in 1869, the Luckenbach brothers who owned the facility rebuilt it immediately as a fine Victorian structure (figure 5) still extant.[31] In 1877 they modernized the operation by adding steam power, followed by an Allis roller mill system in 1882 and a 25,000 bushel grain elevator in 1893.

As the businesses closed, some buildings were converted to tenement dwellings (figure 6). Others housed new enterprises quite different from the original ones, became places of amusement such as a tavern or bowling alley, or fell derelict. Production at the Luckenbach mill gradually switched from flour to feed, and by 1949 all milling had ceased. The once proud Moravian industrial area became more and more neglected and misused. In 1952 the Luckenbach mill was purchased by a dealer in auto parts, and the whole area then became a vast automobile junkyard (figure 7).

7. In the 1950's the old industrial area became an automobile junkyard when a used auto parts dealer bought the former Luckenbach flour mill building for his warehouse.[32]

During the late 1950's and early 1960's the Redevelopment Authority of Bethlehem purchased the land in the old Monocacy industrial area, cleaned it up, and then sold it to the City of Bethlehem. In 1966 a ten-acre portion was leased to Historic Bethlehem Inc., a non-profit, educational organization dedicated to preserving and interpreting Bethlehem's heritage.

Today the eighteenth century industrial area is an outdoor museum of Moravian crafts, trades, and industries. Through extensive archeological and archival research, Historic Bethlehem Inc. has accurately restored the 1761 tannery (figure 8), the 1762 waterworks (figure 27), and the 1869 Luckenbach flour mill. These buildings, as well as twelve other foundations and unrestored structures, are interpreted through tours, exhibits, and special programs to make visitors aware of the legacy of hard work and technological ingenuity left by the Moravian founders of Bethlehem.

8. Today Historic Bethlehem Inc. has restored the old 1761 tannery (above),[33] the 1762 waterworks (figure 27), and the 1869 Luckenbach flour mill (figure 5) and interprets the old Moravian industrial area through exhibits, festivals, guided tours, and educational programs.

MORAVIAN OIL MILL COMPLEX 1745-1814

Linseed oil was an important product in eighteenth century American life. It was widely used in making paints and was considered essential for the preservation of wood. The oil was also burned in lamps, used in the formulation of printing inks, and served as an ingredient for certain medicines.[1]

Linseed oil was obtained from flaxseed, a by-product of the flax crop grown for its fiber. In colonial America flax was harvested after the seeds had matured, so they were easily removed from the stems before preparing the fiber. Only a small portion of the seed recovered was required to sow the next year's crop; the remaining seed was used to make oil.

Flaxseed contains 30 to 40% linseed oil, which can be pressed out only at very high pressures. Since this process requires a large investment in seed pressing machinery and considerable mechanical power, it has been mostly an industrial rather than a home operation.

The early Moravians at Bethlehem raised a large flax crop and were also competent mill builders. Thus it was logical that they would build a mill to produce linseed oil for their own use and for profit by pressing flaxseed for their neighbors.

1745 OIL MILL

Early in 1745, just four years after the founding of Bethlehem, the construction of a new oil mill was recorded in the "Bethlehem Diary":

> February 12. The oil mill was complete enough for our Brethren to make the first linseed oil.
>
> February 18. We had a love feast of oil cakes[2] with the carpenters because they have now completed the oil mill.[3]

This mill was a small log structure 12 × 20 feet (c. 4 × 6 m) in size located in the industrial area along Monocacy Creek.[4] A 1752 map (figure 9) shows the *Oel-Presse* (oil press) building on the eastern side of the tailrace from

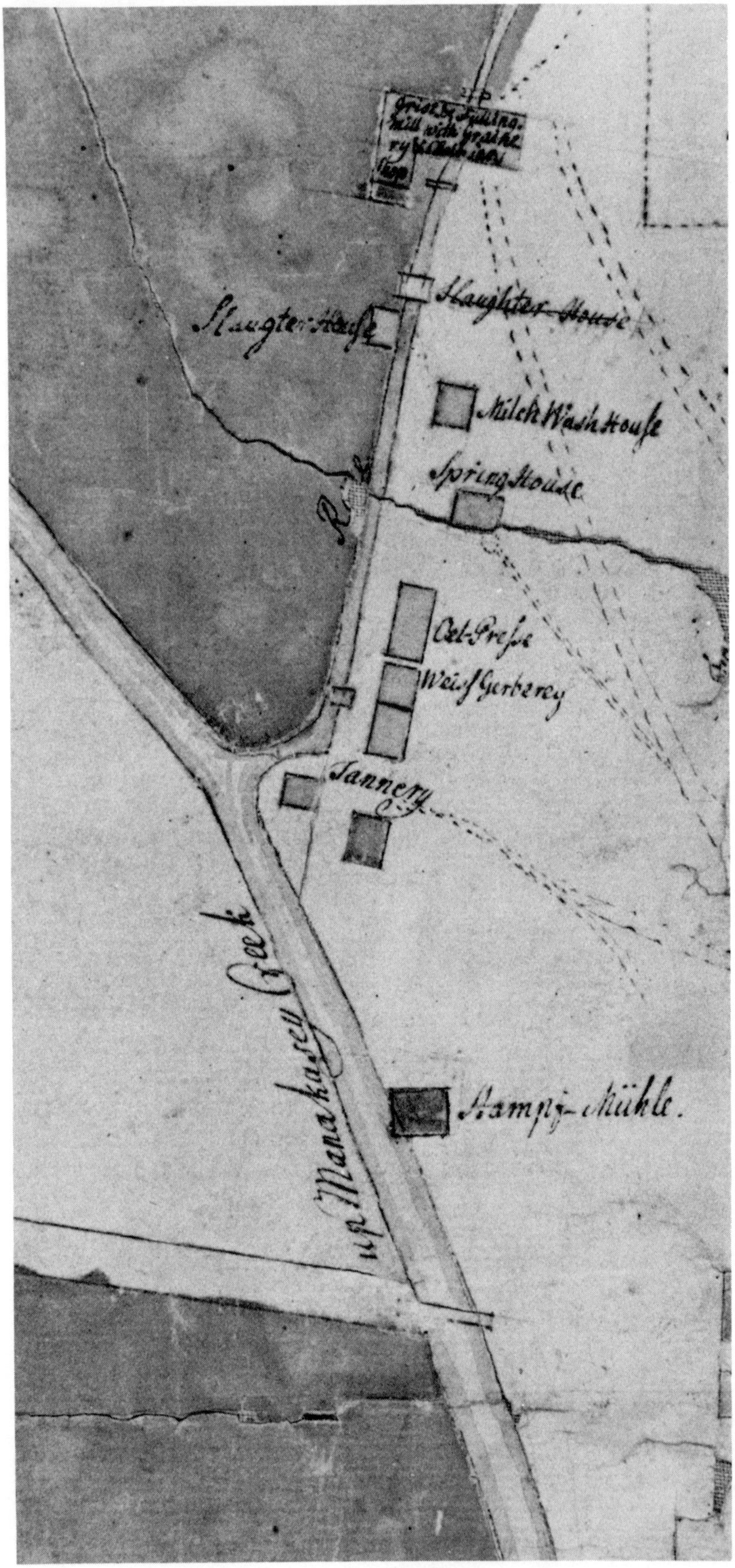

9. A 1752 map of Bethlehem shows both the old 1745 oil mill (Oel-Presse) located on the tailrace of the grist and fulling mill and the new 1752 oil mill (Stampf-Mühle) located on Monocacy Creek.[5] *Note that the new dam and millrace for the 1752 oil mill are absent. This leads us to believe that this map was drawn before construction of the 1752 mill had been completed.*

10. The 1745 oil mill is depicted in this detail from an anonymous 1752 drawing of Bethlehem viewed from the west.[6] No waterwheel is visible, even though the western side of the building faces the millrace (figure 9).

the grist and fulling mill. The western elevation of the building is depicted in a contemporary drawing (figure 10).

Unlike the later Moravian oil mills, oilseed pressing apparently was the only operation in the 1745 oil mill; the tanbark mill is known to have been a separate structure at that time.[4] No details of the machinery used to produce linseed oil from 1745 to 1752 have come to light. The name "oil press" (instead of "oil mill") on the 1752 map, the building's location on the tailrace of another mill, and the lack of a waterwheel in the contemporary drawings (figure 10) suggest that oilseed pressing may not have been a water-powered operation at this location.[7]

Ledger accounts indicate that the 1745 oil mill was quite small, producing c. 200-300 gallons (c. 750-1150 liters) of linseed oil per year for the Moravian economy.[8] However, the annual capacity of the mill must have been several times greater, since the linseed oil pressed for outsiders was not included in these accounts[9] and because oilseed pressing was mainly a wintertime activity (p. 50). George Christ from Nazareth was the oil miller during most of the seven years that this mill was in operation.[10]

During the winter of 1751-1752, both the oil mill and the nearby tanbark mill required numerous repairs.[11] As a result the House Conference of the Single Brethren decided to replace them both with a combined oilseed, tanbark, and hemp mill. Brother John Bishop was assigned to draw up suitable plans,[12] and Hans Christensen was asked to direct its construction.

MILLWRIGHT HANS CHRISTENSEN

Hans Christoph Christensen had arrived in Bethlehem from Europe in September 1751. During the next twenty-five years he became a prolific and innovative mill builder for the Moravian colonists and was a major contributor to their industrial development. He engineered and built both the 1752 and 1765 oil mill buildings at Bethlehem.

Hans Christensen was born in 1716 at Hammelev, Schleswig-Holstein, a farming community in what is now southern Denmark.[13] His father died when he was three years old, and he received a strong religious upbringing from his mother. About 1738 he joined his brother[14] working in the royal grist mill at the nearby town of Haderslev. During his service there his religious convictions deepened. He came in contact with the Moravian Brethren at nearby Stepping and adopted their beliefs. In 1744 Christensen moved to the Moravian center at Herrnhut, east of Dresden, and was formally accepted into their congregation the following year. In December 1745 he was assigned to help build the new Moravian community at Neusalz (now Nowa Sol in western Poland) where he worked as a carpenter for five years. After returning briefly to Herrnhut, he came to Bethlehem in 1751 and served as a millwright in the American Moravian communities for the rest of his life. Christensen never married; he remained a member of the Single Brethren's Choir in

CHRISTENSEN'S MAJOR MILL CONSTRUCTION PROJECTS IN AMERICA[18]

For Moravian Communities

1751	Grist mill expansion	Friedensthal PA
1752	Oilseed, hemp, and tanbark mill	Bethlehem PA
1754	First waterworks	Bethlehem PA
1754-56	Grist and saw mill	Bethabara NC
1757-58	Grist and saw mill	Lititz PA
1761-62	New waterworks	Bethlehem PA
1764	Tanbark mill	Bethlehem PA[19]
1764	Grist mill	Christians-brunn PA[20]
1764-68	Oilseed, tanbark, snuff, hemp, and groat mill	Bethlehem PA
1765	Fulling mill	Lititz PA[21]
1770-71	Grist mill	Hope NJ
1771	Rebuilt fulling mill	Bethlehem PA[22]
1776	Grist mill	Lititz PA

For Private Owners

1767	Grist(?) mill	Allentown PA[23]
1767	Hemp mill	Macungie PA[24]
1769	Grist(?) mill	Easton PA[25]
1774	Grist(?) mill	Reading PA[26]
1774	Saw mill	Unknown[27]
1774	Grist mill	Oley Valley PA[28]

Bethlehem until his death. He died of consumption in 1776 at Lititz, Pennsylvania while rebuilding a burned grist mill.

Christensen is best known for designing and building the Bethlehem waterworks, the first pumped municipal water supply system in America and one of his most innovative engineering projects (p. 14). However, a broader picture of his talents emerges when one lists his major mill construction projects during the twenty-five years he was in America (see box). He built mills for many different processes: grinding grain, sawing wood, stamping and pressing oilseeds, stamping and rolling hemp, stamping tanbark, and pumping water. Christensen was clearly a versatile and prolific millwright who could construct many types of large mechanical equipment. His talents were also appreciated outside the Moravian community, and he frequently contracted to assist non-Moravians with mill construction projects.

It is not exactly clear where Christensen acquired his expertise as a millwright. All we know about his European experience is that he worked five or six years in a grist mill at Haderslev and that he worked as a carpenter at Neusalz for five years. He may have been involved in building or repairing mills while at Neusalz; for upon his arrival in America in 1751, Christensen was immediately assigned to install a new run of millstones at the Friedensthal grist mill. However, he comments in his memoir that "[The Friedensthal mill] was the beginning of my subsequent mill construction work"; so he was still learning his trade at that time. From 1751 to 1762 Christensen was one of several builders involved in Moravian mill construction projects in America.[15] By 1762 his reputation as a skilled millwright was firmly established, and he was formally in charge of all mill construction work by that time.[16]

One of Christensen's contemporaries left this brief comment on his personality:

> By nature he had a somewhat rough and authoritative manner, which sometimes made it difficult for those who had contact with him or worked under his supervision. Halfheartedness and apparent unfaithfulness anywhere angered him very much. However, everyone will have to agree that he acted out of a faithful mind and heart.[17]

Christensen was clearly the type of man who brooked no interference with "getting on with the job," one good reason for his success as a millwright.

1752 OIL MILL

The site selected for the new oilseed, tanbark, and hemp mill was on the eastern side of Monocacy Creek about 200 feet (61 m) south of the 1745 oil mill (figure 9). During the summer of 1752 Christensen directed the construction of the new multipurpose mill.[13] The building measured 34 × 46 feet (10.4 × 14.0 m) and was of wooden frame construction.[29] The Moravians built a new dam across the Monocacy and a new raceway to bring water to the mill.[30] No detailed records describing the construction of the 1752 oil mill have survived. Three contemporary drawings of the entire town of Bethlehem provide small images of the building (figure 11), but unfortunately they differ greatly in architectural detail. Two of the views show an undershot waterwheel beneath the roof (or lean-to shelter) on the western side of the mill, but the third has none in sight even though the building is viewed from the same direction.

This new "stamping mill" (as it was frequently called) housed machinery for processing oilseeds, tanbark, and hemp. Although no descriptions of the actual equipment used have been found, the name "stamping mill" suggests machinery of the type installed in the subsequent 1765 oil mill: oilseed stampers, a wedge press, tanbark stampers, and hemp stampers (figures 29-31). Presumably all three

11. During the 1750's, three artists drew town views of Bethlehem that included the 1752 oil mill: Nicholas Garrison Jr.[31] (top), an anonymous artist[32] (center), and Thomas Pownall[33] (bottom). Their three renditions of the building differ greatly in architectural detail, leaving us uncertain about what the mill actually looked like. Both the Garrison and Pownall versions were redrawn by London engravers, who might have interpreted the originals as they saw fit.

operations were powered by the same waterwheel but were not used simultaneously. When Bethlehem's first municipal water pump was built in 1754, it was temporarily installed at the 1752 oil mill for testing before a permanent waterworks building was constructed.[34]

George Christ operated the new 1752 oil mill at its startup.[35] Albrecht Klotz, with his Black assistant Christian Anton, took over in 1757.[36] With the reorganization of the Moravian economy in 1762, the oil mill became part of the tannery business; and the tanner John George Geitner supervised its operation.[37]

The 1752 oil mill produced c. 750-1550 gallons (c. 2800-5900 liters) of linseed oil annually for Moravian use,[38] four or five times the production of the 1745 mill. The annual capacity of the 1752 mill was undoubtedly larger than these figures indicate, however, since oilseed pressing was also done for outside customers and the equipment was often idle during the summer months (p. 50). In 1752-1753, 137 gallons (519 liters) of rapeseed oil was produced,[39] apparently an experiment that was not repeated. No production records have been found for the hemp and tanbark stamped at the 1752 mill.

Linseed oil became an important item of trade with other colonists, because it provided cash to purchase the commodities that could not be produced in Bethlehem. In 1754 the Moravians attempted to use the Lehigh and Delaware Rivers for transporting their linseed oil to Philadelphia. They built a flatboat, the "Little Irene," which was launched amid general rejoicing and music. This boat carried a cargo of linseed oil to Philadelphia where it was successfully marketed. However, the crew found it impossible to navigate the boat up the rapids of the Delaware River for the return trip, and the "Little Irene" had to be sold in Philadelphia.[40] Thereafter all linseed oil shipments went by land.

Then on November 18, 1763 came a major industrial and economic disaster for the Bethlehem community:

> This morning during the first hour past midnight a fire broke out in the oil mill. The wind was blowing away from the community, it was snowing, and there was highly combustible material. Hence the fire progressed so quickly that the guards were not aware of it until the flames had already affected the nearby waterworks.[41]

The fire occurred only the day after rumors had been heard that hostile Indians were in the area. After the initial confusion, only part of the Moravian Brethren were assigned to fight the fire while the others patrolled the town in order to guard against ambush.[42] The oil mill building was beyond saving by then, and the firefighters concentrated on rescuing the nearby waterworks.

The fire could not have come at a worse time for the Moravian community. Bethlehem was overcrowded with refugees from the ongoing French and Indian War, and relations with non-Moravian neighbors had become very strained. Even during times of peace, Moravian missionary efforts among the Indians were regarded with a certain wariness by nearby white settlers. Now during a time of war, all Indians were considered hostile. On the day of the oil mill fire, news had reached Bethlehem that their Indian mission at Wechquetank had been burned to the ground.[43] Then footprints were found leading away from the burned oil mill towards a neighboring white settlement. There was no doubt in the minds of the Moravians that the oil mill fire was the result of arson.[44]

The small Moravian community at Bethlehem had a tradition of rallying to meet an adversity, and the oil mill fire was no exception. Even in the face of disaster, they maintained their positive attitude toward life:

> The loss we suffered was substantial, but we have cause to thank our Lord with all our hearts. Not one single Brother was injured while fighting the fire. And a persistent strong wind came from the northeast, thus keeping the fire away from the town and preventing exposure to a far greater danger.[42]

Rather than complaining about the disastrous fire, the Moravians quickly turned to planning how they would rebuild their lost oil mill. John Arbo, a newcomer to Bethlehem, was selected to oversee the project; and Hans Christensen again served as the millwright.

12. Portrait of John Arbo painted by Valentine Haidt in the 1760's.[48]

ADMINISTRATOR JOHN ARBO

Two years before the disastrous oil mill fire, a new administrator had arrived from Germany to manage the Single Brethren's Choir in Bethlehem. This was John Arbo,[45] a former mathematics teacher with a talent for financial affairs.

John Arbo was born in 1713 in the small German town of Sörup in Schleswig-Holstein.[46] At the age of seven he contracted smallpox, and the disease left him crippled and unable to do physical work.[47] He received considerable formal education and was appointed a village schoolmaster (location unknown) in 1731 at the age of eighteen. In 1737 he joined the teaching staff of the German Writing and Arithmetic School in Flensburg, Schleswig-Holstein.

During his twenties Arbo was active with a number of different religious groups, continually searching for beliefs that he could adopt as his own. Finally in 1742 at the age of twenty-nine he traveled to Herrnhut and joined the Moravian Brethren with whom he found the religious fulfillment he had been seeking. For the next fourteen years he served as a teacher in church schools at Lindheim, Marienborn, Hennersdorf, and Niesky in Germany. At the same time, his experience in various minor church positions helped develop his managerial abilities, and in 1756 he left teaching to become Registrar and Secretary to the Administrative Office of the church in Herrnhut. Three years later he was appointed an assistant to the Warden of the Single Brethren's Choir there and became responsible for keeping all the financial records of that group. In 1761 he was ordained a Deacon in the Moravian Church and assigned to Pennsylvania.

Shortly after his arrival in Bethlehem, Arbo became the Warden[49] of the Single Brethren's Choir and helped supervise its financial reorganization. In 1762 Bethlehem changed from a church-directed communal economy to a new economic system having both individual and congregational enterprises.[50] With this change the Single Brethren's Choir became an independent financial unit operating specific trades (shoemaking, hatmaking, saddlemaking, etc.) and industries (hemp mill, brewery, etc.). It bought raw materials, paid wages to its members for processing them, sold finished products, and kept regular business records. Arbo successfully administered this difficult transition and helped establish a profitable operation for the Single Brethren. His next major project was planning and supervising the rebuilding of the Bethlehem oil mill from 1764 to 1768. He continued to serve as Warden of the Single Brethren, a member of the *Aufseher Collegium*, and in various other church administrative posts until his death in 1772 at the age of fifty-nine.

John Arbo was quite a scholarly man. Dozens of books from his library are now in the collection of the Moravian Archives in Bethlehem.[51] They indicate his keen interest in a wide variety of subjects: mathematics, medicine, mill construction, mineralogy, glassmaking, history, geography, and religion. His methodical study habits are revealed in several unique books that Arbo himself produced. One manuscript volume contains a neat, elaborate compilation of mathematical principles and their applications.[52] A world atlas of 80 maps appears to have been personally assembled by Arbo from individual maps collected over a period of years; a detailed Table of Contents in his handwriting appears on the front sheet of this volume.[53] He was elected to membership in the American Philosophical Society of Philadelphia in 1768[54] and later reported to them about his experiments at the Bethlehem oil mill (figure 13).

P H I L A D E L P H I A.

THE AMERICAN PHILOSOPHICAL SOCIETY, held at Philadelphia, *for promoting useful Knowledge*, have been favoured with the following Communications, *to wit*.

THE *upper and under Delineation* of 28 different *Butterflies of the Island of Jamaica*; by Mr. SAMUEL FELSTED.

A Paper tending to *disprove* the Opinion of *Snakes being capable to charm Birds*; by DAVID RITTENHOUSE, A. M.

A Letter from Mr. JOHN ARBO, of Bethlehem, to Dr. Thomas Bond, containing a farther Account of the Culture of *Sunflowers* for extracting *Oil*---Remarks on *Currants*, and a Receipt for making *Currant Wine*---Various Specimens of *Oil*, extracted from *Sun-flower Seed*, *Peach Kernels*, *Cabbage Seed*, *Walnuts*, *Hickory-nuts*, *Poppy Seed*, *Laurel Seed*, *Pompion Seed*, *Sallad Seed*, and *Cotton Seed*. A Method to destroy *Caterpillars*---and some Samples of *Earth*, from which, he thinks, *good yellow Oaker* might be made.

13. In 1771 Arbo sent a report to the American Philosophical Society in Philadelphia describing his experiments at the Bethlehem oil mill extracting fatty oils from unusual seeds.[55]

A brief glimpse of Arbo's personality is provided by the minister who completed his memoir after his death:

> By nature he had an active and quick demeanor which often offended one or the other of the Brethren. However, they were soon convinced of his good intentions and that he was only concerned that everything was done according to the will of Christ. His willingness to be of service to everyone made him popular with outsiders.[46]

Thus one might picture John Arbo during his Bethlehem years as a practical, thorough, hard-driving, but somewhat volatile administrator whose accomplishments demonstrate his effectiveness. His meticulous records and justifications of his actions[56] seem to indicate that he also worried a good deal about what his contemporaries and successors might think of him.

As we shall see, it was the combination of Arbo's administrative and financial abilities with Christensen's practical experience as a millwright and builder that assured the successful construction of a new and much larger oil mill for Bethlehem. In many respects the two men had similar personalities, although they differed greatly in educational background. Both hated inefficiency with a passion. Both worked very hard and expected their fellow workers to do the same. They had trouble understanding the incompetent, the lazy, and the dishonorable and constantly fought against such "un-Christian" traits. United by a common purpose and complementary talents, they started planning the new oil mill project in 1764.

BUILDING THE 1765 OIL MILL

The new oil mill was a far more ambitious undertaking than the old one. The fire had destroyed a wooden structure; the new mill was to be built of stone. The 1752 oil mill had processed oilseeds, hemp, and tanbark using the power of one waterwheel. That same equipment was to be rebuilt in the new larger mill, but additional machinery for grinding groats (coarsely ground cereal grains such as pearl barley) and for fulling leather would also be installed. Two waterwheels would now be required to power all this machinery. To call this new building an "oil mill" was really a misnomer; it was a complex, multipurpose mill, the most versatile manufacturing facility built in eighteenth century Bethlehem.

In 1764 church administrators decided that the Single Brethren's Choir would build and manage the new oil mill.[57] But the tanner, John George Geitner, who was not one of the Single Brethren, maintained that he should operate the tanbark stamping and leather fulling machinery since they were an intimate part of his tanning operations. Arbo considered such a division of authority to be both economically and politically undesirable.[58] Placing separately-supervised industries together in the new mill could easily lead to conflicts, especially when there was insufficient waterpower to operate all the machinery simultaneously.

Arbo proposed that a separate tanbark and leather fulling mill be built for the tannery or (should that prove impossible) that the Single Brethren should run the entire new oil mill facility but charge the tannery only a modest price for the work done. Geitner objected, saying that he needed immediate access to the fulling mill whenever he removed leather from the tanning vats. Moreover one of Geitner's employees, Andrew the Black, was already involved in preparing tanbark for the tannery, and he would not be fully employed if the Single Brethren now did this work. Extensive negotiations followed with the church administration mediating between Arbo and Geitner. It was finally agreed that the tannery should pay £25 per year to the Single Brethren's Choir for running the tanbark and leather fulling mills at the new building and that Andrew would work for the Single Brethren rather than the tannery.[58]

In December 1764 Christensen drew up the plan for the new oil mill building.[59] A few pieces of equipment (a waterwheel, some ironwork, etc.) had been saved from the fire at the 1752 oil mill, and these were used again in the new mill.[60] During the winter Christensen and his workers

14. Contemporary drawing of the completed 1765 Bethlehem oil mill viewed from the southwest (tail-race side).[62] Note the "1765" banner on the flag pole.

Notice is hereby given to the PUBLIC,

THAT at Bethlehem, in Northampton County, the Mills for rubbing Hemp, ſtamping Millet, and preſſing Oil, are done and ſet at work already. The Hemp is not rubbed with a Pumiceſtone in the common Way, that being attended with many Dangers; but it is ſtamped in a particular Manner, and becomes pliabler and fitter than with the Stone. In the Oil Mill every one may alſo have his Flaxſeed cleanſed. The Mills for Groats, peeled Barley, Oatmeal and Spelt, will alſo be finiſhed this Summer, and ſet to work in July or Auguſt next. The further Particulars may be enquired at the ſaid Bethlehem, by

JOHN ARBO.

15. In March 1766 John Arbo placed this advertisement in the Pennsylvania "Gazette" announcing that the new Bethlehem oil mill was open for public business.[63]

cut stones and prepared lumber for the new building. In June 1765 they dug a diversion channel for the Monocacy Creek and built a temporary dam to divert water from the building site. Finally on July 12 the cornerstone of the new oil mill building was laid. It required 22 days to install the stone foundation and build the new mill race; and during this work the site was *manually* pumped day and night to remove the water seeping into the trenches. Construction work then proceeded quite rapidly. By the end of September the walls were two stories high, and a new dam across Monocacy Creek was nearly finished. When the rafters were raised on October 9th, the Bethlehem diarist boasted, "This will be one of the most solid and durable buildings and the only one of its kind to be found in this land."[61] By the end of November the building was nearly complete. A drawing of the new mill was made (figure 14) and proudly sent to the Moravian Church headquarters in Herrnhut, Germany.

Work on the complex machinery inside the mill proceeded as the building was being completed. By early December both waterwheels and many of the gears and drive shafts had been installed. An announcement was posted informing the public about the coming opening of the new mill at Bethlehem (figure 58). On December 27 the hemp stamping mill was put into operation, no doubt with considerable ceremony; and the first day's customers had their hemp stamped free of charge. Flaxseed stamping began on February 18, 1766, and the first linseed oil was pressed the following day. The tanbark mill started up on March 3rd. By then the operation was going so well that Arbo placed advertisements in Pennsylvania newspapers to attract more non-Moravian customers (figure 15). The groat milling machinery was completed early in the fall of 1766, and the production of barley groats and oat meal began.

Late in 1767 a water-powered hemp roller mill was installed on the second floor to soften hemp fiber for weaving into cloth. That same winter a separate drying kiln was built 40 feet (12 m) south of the oil mill (figure 3) to dry oats and snuff tobacco prior to processing.[64] The planned leather fulling operation, however, was apparently never realized.[65]

John Arbo kept meticulous records throughout the entire oil mill construction project. He recorded the day-by-day activities in an "Oil Mill Diary" and maintained a detailed list of expenses in an "Oil Mill Account Book." Fortunately these records have survived; and complete English translations are presented in a later chapter (pp. 76-99). These fascinating documents describe in matter-of-fact style the tremendous amount of organizational effort, technical expertise, and physical labor required to construct the 1765 Bethlehem oil mill with all its complex machinery. They also reveal interesting details about work practices of the period; for example, one of the expenses of building the hemp roller mill was ten quarts of rum for the workers.[66]

In addition to these two books, a set of machinery drawings for the new oil mill was made in the summer of 1766 to send to the church authorities in Herrnhut. These remarkable plans have also survived (figures 28-31) showing exactly what types of machinery were built inside the oil mill and complementing the detailed diary and financial accounts.

SINGLE BRETHREN OPERATION 1765-1814

The new oil mill with its many products rapidly became a major source of income for the Single Brethren's Choir. From 1767 through 1787 net profits averaged £155 per year (figure 16) making it one of the more profitable industries of Bethlehem. No doubt Arbo's firm administrative hand directing the Single Brethren (until

16. Annual profit or loss on the oil mill industries under the management of the Single Brethren's Choir from 1767 to 1814.[67]

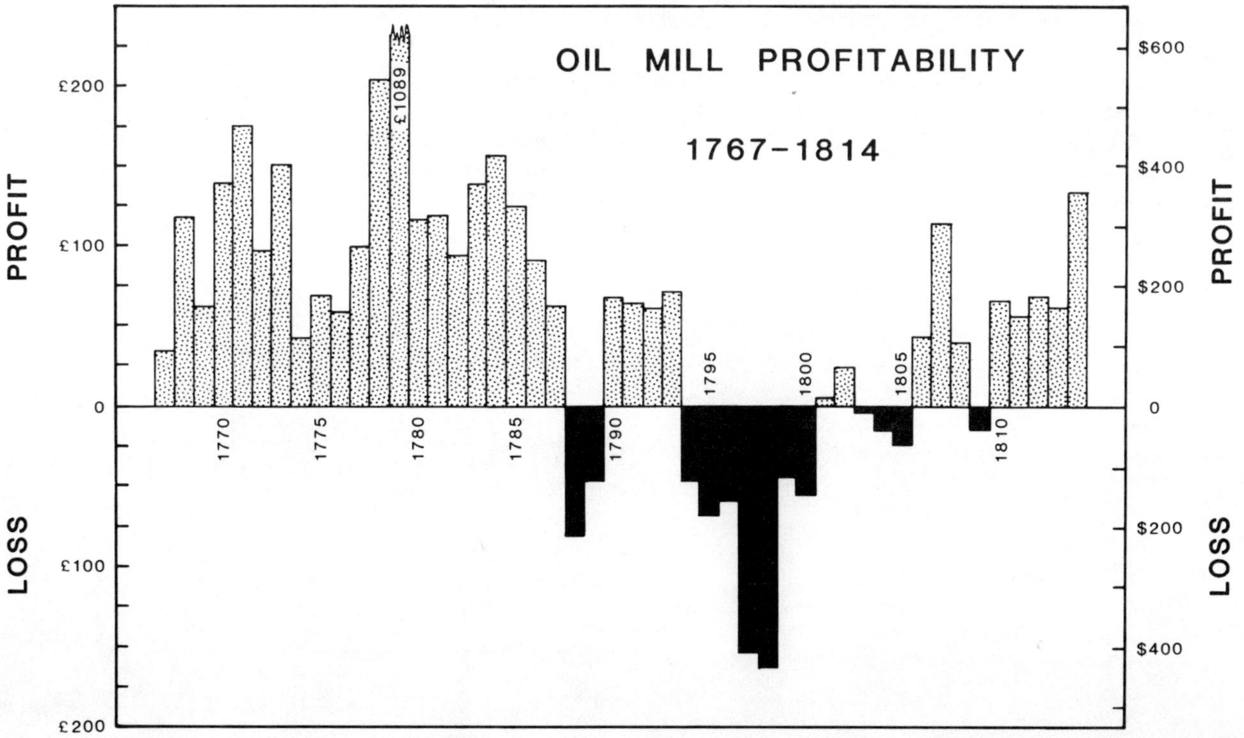

his death in 1772) and the increased demand for agricultural products during the Revolutionary War (1776-1783) contributed to the profitability of the operation.

Linseed oil, barley groats, and oat meal were the principal cash-earning products during this period.[68] As the years progressed, however, additional products were manufactured at the oil mill. Buckwheat was hulled and then ground into groats and meal. Millet was stamped for use in millet porridge.[69] Barley malt was ground for the local brewery,[70] and split peas were produced.[71] From 1766 to 1793 the oil mill dried, stamped, and sifted tobacco to be used in making snuff.[72] In 1794 a proper snuff mill was installed in the building for this purpose.[73]

The 1765 oil mill complex became widely recognized as an exceptional engineering achievement, and it was proudly shown to the frequent visitors who came to town. Future president John Adams visited Bethlehem in April 1777 and remarked in a letter home to his wife about the impressive oil mill he had seen.[74] General George Washington stayed overnight in Bethlehem in July 1782 and was shown the oil mill during his tour of the community.[75] Prominent European visitors such as the German scientist Johann Schoepf[76] and the French social reformer Duke de La Rochefoucault[77] also wrote about seeing the Bethlehem oil mill during their travels in America.

During the late 1780's the profitability of the oil mill industries started to decline and never returned to the earlier levels. From 1788 through 1814 there were net losses in 13 out of the 27 years (figure 16). This coincided with frequent losses in many of the other trades and industries operated by the Single Brethren's Choir. Their membership had become older and had declined both in number and in competence.[78] Although the oil mill showed a profit once again in 1810-1814, operational deficits at the Single Brethren's Choir continued to mount. Church administrators reluctantly decided to disband the group, transfer its property back to the General Diacony, and lease out its industries for private operation.[79] Assuming a tenant could be found, leasing the oil mill would assure a regular income from the Moravian Church's investment in the building. Thus on May 31, 1814, Single Brethren operation of the oil mill industries came to an end.[80]

17. This detail from Garrison and Oerter's 1784 view of Bethlehem (figure 4) indicates that a shed was added onto the western side of the 1765 oil mill several years after it was built.

LOWER GRIST MILL AND WATERWORKS 1814-1934

The end of the Single Brethren's operation in 1814 brought major changes in the role of the Bethlehem oil mill. Two types of industry occupied the building during the next hundred years. First came a succession of tenant millers who used the existing groat mill equipment to develop a flourishing grist milling business. Then in 1832 the town of Bethlehem relocated its waterworks into the building. Both operations shared the facility until 1875 when the grist mill was shut down because all the available power was needed for the waterworks. As usage changed, the popular name of the building changed too, from "oil mill" to "lower grist mill" to "Bethlehem buckwheat mill" to "waterworks."

18. Samuel Schultz (Schulz) was the tenant miller from 1823 to 1835.[8] When the new town water pump was installed at the oil mill in 1832, he was required to operate and maintain it as part of his lease.

TENANT MILLERS 1814-1875

In 1814 the mill was rented from the Moravian Church by Gideon Helwig,[1] a former cook.[2] He built a residence for himself and his family in the western half of the building where the old tanbark and hemp milling machinery had been.[3] During his eight-year tenancy, Helwig ground oat meal, barley groats, buckwheat meal, brewery malt, pig feed, corn, and snuff.[4] New equipment for drying oats and other grains prior to grinding was probably installed soon after he moved in.[5] He also sold large quantities of candles which he made at the mill,[6] possibly melting the tallow on the stove formerly used to roast the crushed flaxseed for the oilseed mill. Helwig operated the mill until his death in 1822. Samuel Schultz[7] (figure 18) then took over the milling operation and ran it until 1835.[9] He was the tenant miller in 1832 when Bethlehem's water pumping operation was moved into the ground floor of the mill. Thus he was the first miller to be required, as part of his lease, to operate and maintain the community's water pump.[10]

In 1835 Charles D. Bishop[11] (figure 19), who had briefly managed the oil mill in 1811, became the resident miller. "It was through the care and excellent management of . . . Charles D. Bishop" that the high reputation of Bethlehem buckwheat flour and oat meal was established, "he being the first to make the meal to perfection during the time he was lessee from 1835 to 1847."[13] Bishop's interests were not limited to milling. He was also a director of the Bethlehem Water Company,[14] a member of the influential *Aufseher Collegium,*[10] and the first manager of the Eagle Hotel in Bethlehem.[15]

The lower grist mill was again for rent in 1847 (figure 21). It was leased to Francis Zoller[16] until 1854,[17] but the exact length of his tenancy is unknown. In 1860 there was mention of a Bethlehem buckwheat miller named L. Snyder,[18] who may have been the oil mill tenant at that time. The Bishop name reappeared in 1863 when Gilbert Bishop[19] (figure 20), the son of former miller Charles D. Bishop, acquired the lease. He operated a successful buckwheat and oat meal milling operation at the oil mill for the next eleven years.

Since the grist mill and the waterworks shared the same marginally-adequate source of waterpower, all the tenant millers had to adjust their grinding operations to the growing water needs of the town. As early as 1819, the *Aufseher Collegium* asked miller Helwig to stop using water from the Monocacy Creek for two days each week so that the adjacent waterworks could keep the town's reservoir full.[21] The miller could grind only after the community's thirst had been satisfied!

When the waterworks was moved to the oil mill building in 1832, both the grist mill and the water pump were powered by the same waterwheel. Since there was insufficient power to run both the mill and pump simultaneously,[22] this situation inevitably led to conflicts. In 1834 the *Aufseher Collegium* agreed to compensate miller Schultz for the injury to his milling business caused by the installation of the new water pump.[10] And when C. D. Bishop succeeded Schultz as resident miller, the rent on the building was reduced from $160 to $125 per year.[23] Even with the installation of steam pumps in 1868, the 1832 water-powered pump was still kept as a backup system and frequently used (p. 67). By early 1874 Bethlehem's increasing demands for water required that the waterwheel run almost continuously to operate the pump, and Gilbert Bishop had little power available for grinding grain.[24] This, together with a three-year delay in payment of a $235.75 debt owed him by the borough,[25] apparently caused him to terminate his lease and move his business elsewhere.[26]

The installation of a new steam pump for the waterworks in 1874 freed some of the waterpower for other duty. Lewis Snyder,[27] who had tried for several years to rent the mill,[28] took possession in April and started grinding oat meal and buckwheat flour. However, his lease agreement included

> The express understanding that the water wheel pump work at least 8 out of every 24 hours, always in the night, but in case of a fire to start the water

19. Charles D. Bishop, the resident miller from 1835 to 1847,[12] established a wide reputation for the quality buckwheat flour and oat meal produced at the lower grist mill. He also served as a director of the Bethlehem Water Company.

20. Gilbert Bishop, son of former miller Charles D. Bishop, also produced buckwheat flour and oat meal at the lower grist mill from 1863 to 1874.[20]

21. *This bilingual 1846 broadside announced the availability of the lower grist mill facilities being vacated by Charles D. Bishop.*[14] *The oil mill building was known as the "lower" grist mill at that time to distinguish it from the 1751 "old Moravian" grist mill located 400 feet (122 m) upstream.*

> wheel pump at once, and to have the same going as long as required, . . . and in case anything should happen to the steam pump at any time to set the water wheel in motion, and if not required to go more that twice in 24 hours, he is not to have anything for his labor or the use of the water power.[29]

Snyder remained as tenant miller for less than one year. In 1875 the Water Board terminated his lease and then renovated the miller's residence for rental to the waterworks engineer.[30] The equipment remained, but the lower grist mill stones were permanently silenced.

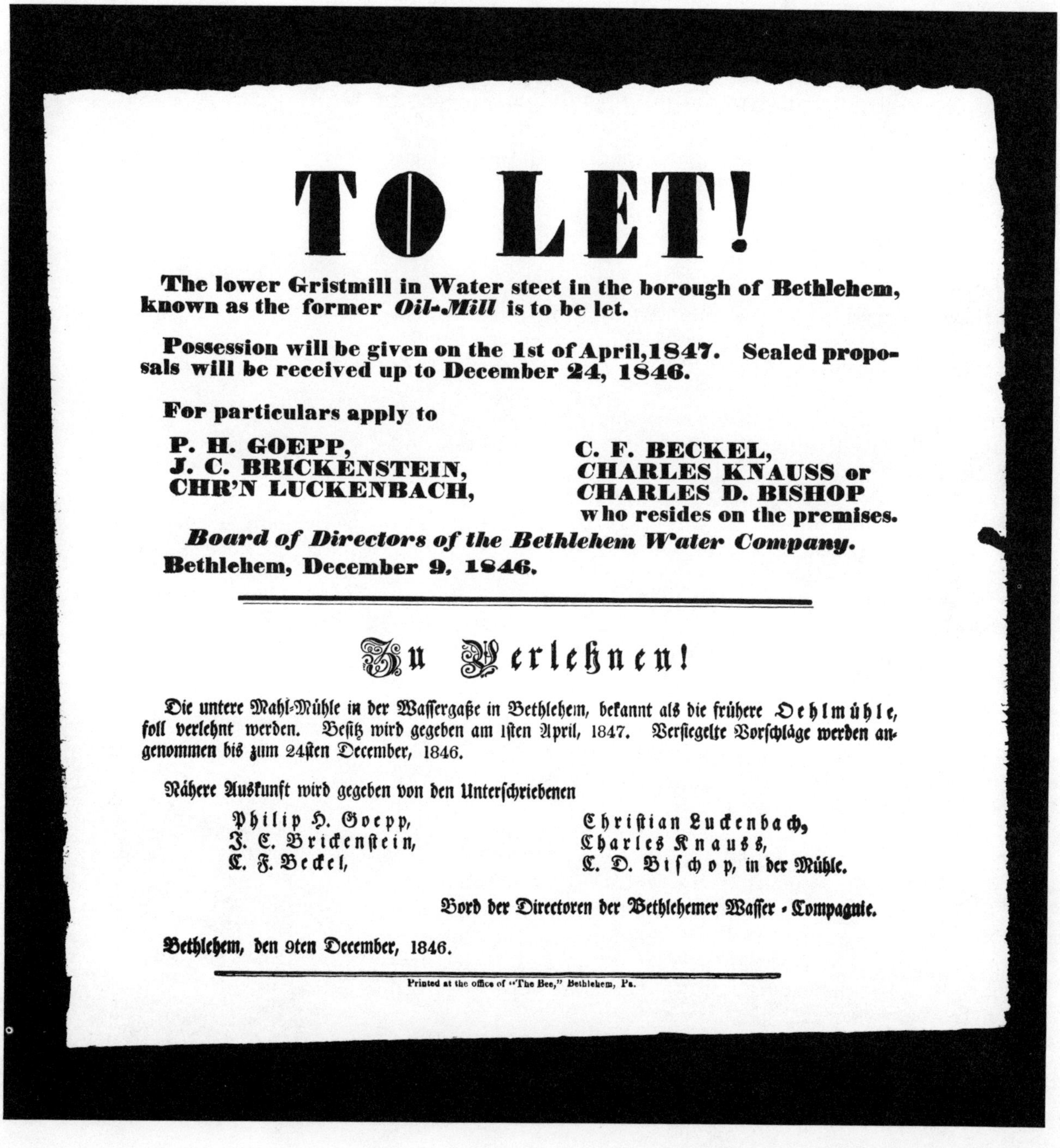

TO LET!

The lower Gristmill in Water steet in the borough of Bethlehem, known as the former *Oil-Mill* is to be let.

Possession will be given on the 1st of April, 1847. Sealed proposals will be received up to December 24, 1846.

For particulars apply to

P. H. GOEPP,	**C. F. BECKEL,**
J. C. BRICKENSTEIN,	***CHARLES KNAUSS* or**
CHR'N LUCKENBACH,	***CHARLES* D. BISHOP**
	who resides on the premises.

Board of Directors of the Bethlehem Water Company.

Bethlehem, December 9, 1846.

Zu Verlehnen!

Die untere Mahl-Mühle in der Wassergaße in Bethlehem, bekannt als die frühere Oehlmühle, soll verlehnt werden. Besitz wird gegeben am 1sten April, 1847. Versiegelte Vorschläge werden angenommen bis zum 24sten December, 1846.

Nähere Auskunft wird gegeben von den Unterschriebenen

Philip H. Goepp,	Christian Luckenbach,
J. C. Brickenstein,	Charles Knauss,
C. F. Beckel,	C. D. Bischop, in der Mühle.

Bord der Directoren der Bethlehemer Wasser-Compagnie.

Bethlehem, den 9ten December, 1846.

Printed at the office of "The Bee," Bethlehem, Pa.

WATERWORKS 1832-1913

In 1832 the oil mill building began the third phase of its history, serving as the Bethlehem waterworks. It succeeded two earlier water-pumping installations designed and built by Hans Christensen in 1754 and 1762 at a site just east of the oil mill.[31] Water was drawn from a bountiful spring located in the center of the Monocacy industrial area.[32]

By the 1830's the increasing demand for water dictated that a new works be built, and the oil mill was the most convenient site. It had extra space, available waterpower, and was located only 20 feet (6 m) west of the existing waterworks so only minimal piping changes would be required. In 1832 Charles D. Bishop, representing the Bethlehem Water Company, contracted with Rush and Muhlenberg of Philadelphia for a water pump, drive shaft, and auxiliary equipment.[33] These were installed by Bishop and Ernst Gehbe (the superintendent of the waterworks) in the eastern half of the building where the former oilseed mill had been located. Overhauled and rebored in 1874,[34] this dependable water-powered pump remained in service until 1888.

As a result of this relocation of the waterworks operation, ownership of the oil mill building was transferred from the Moravian General Diacony to a church-organized local civil authority (the *Bürgercasse*) in 1832-1834.[35] In 1845 the Bethlehem Water Company was incorporated under a state charter and acquired title to the oil mill building along with its grist milling and pumping machinery, the nearby spring, and the entire water distribution system.[36] Subscribers to this private water service paid an annual "water tax" of $10 to $11. Demands for expansion of the system, which the Water Company refused to finance, and complaints about insufficient water led the municipal authorities to consider purchasing the company.[37] After unsuccessful negotiations in 1856-1857 and 1866-1867,[38] matters came to a head in 1871 when a corner of the reservoir collapsed, cutting off the town's water supply. The Water Company was "unwilling to incur the additional expense and responsibility of erecting a new reservoir,"[39] thus forcing borough authorities to act. Consequently the Borough of Bethlehem purchased the Bethlehem Water Company in May 1871 for c. $25,000, and the oil mill became a municipal property.[40]

During the 81 years that the waterworks was located at the oil mill, there was a continual race to keep up with the water demands of the local residents. Bethlehem's population grew from c. 650 in 1830 to 12,837 in 1910,[41] and projects to increase pumping capacity were required in nearly every decade.

The first changes focused on improving the waterwheel driving the pump. The Monocacy Creek had always been subject to flash flooding and high water for days at a time.[42] The resulting backwater in the oil mill wheelpit

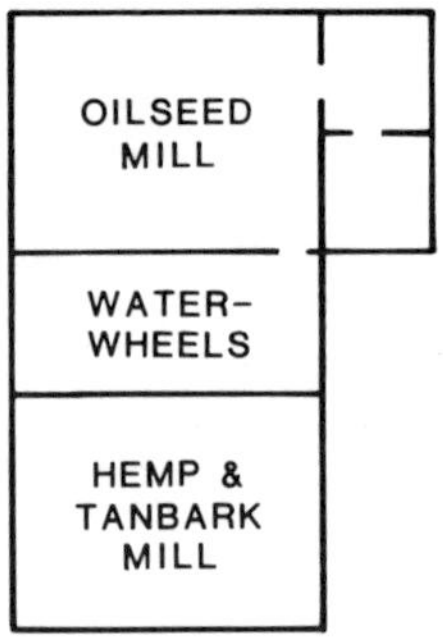

1765

22. This detail from an 1873 map entitled "Birds-Eye View of the Bethlehems" shows the oil mill from the southwest.[51] The chimney for the 1868 steam boilers now dominates the southern side of the building.

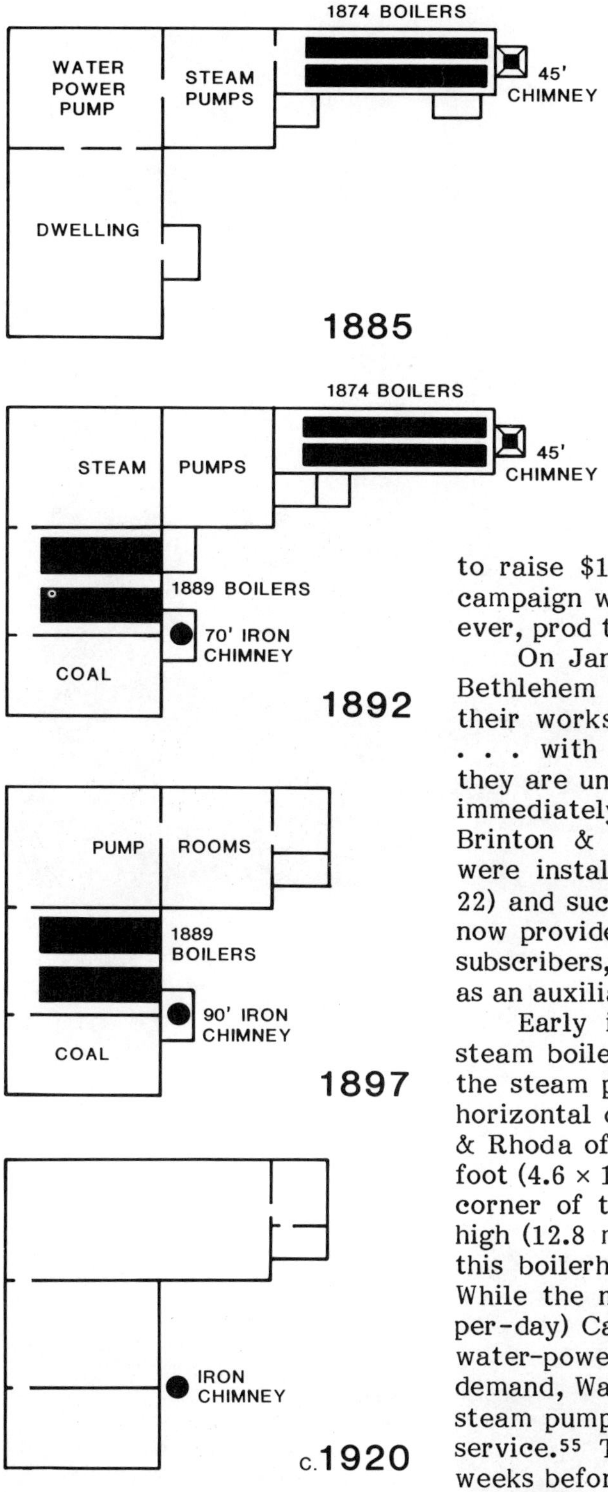

23. Insurance maps for the Borough of Bethlehem show how the floor plan of the original 1765 oil mill was progressively changed as the 1874 and 1889 boiler systems were installed and then later removed.[53]

would stop the undershot waterwheel and shut down the pump.[43] In 1846 the waterwheel was replaced with a Howd patent turbine,[44] probably because a turbine is less subject to backwater shutdown. The performance of the turbine apparently was disappointing, for ten years later it was replaced by a new undershot waterwheel to gain additional power and pumping capacity.[45]

By 1867 there were repeated public outcries about the lack of adequate water in Bethlehem.[42] Since the reservoir could hold less than one day's supply, the situation became critical. A planning committee was formed,[46] and a Philadelphia hydraulic engineer, H. P. Birkenbine, was consulted.[47] A group of concerned citizens attempted to raise $125,000 for a new public waterworks, but the campaign was a failure.[48] Community pressure did, however, prod the Water Company into action.

On January 6, 1868 it was announced that "the old Bethlehem Water Company will put a steam engine in their works, so as to be able to furnish . . . Bethlehem . . . with water when the Monocacy is swollen, when they are unable to do so by water power."[49] Work began immediately. Two boilers and twin steam pumps from the Brinton & Henderson Hydraulic Works of Philadelphia were installed in the eastern half of the oil mill (figure 22) and successfully tested on June 8th.[50] Steam power now provided a satisfactory water supply to all company subscribers, and the old water-powered pump was retained as an auxiliary system.

Early in 1874 it was suddenly discovered that the steam boilers were so corroded as to be unsafe and that the steam pumps were no longer serviceable.[52] Two new horizontal cylindrical boilers were purchased from Noble & Rhoda of Allentown and were installed in a new 15 × 53 foot (4.6 × 16.2 m) brick boilerhouse built at the southeast corner of the old stone oil mill (figure 23). A 45-foot-high (12.8 m) brick chimney erected at the south end of this boilerhouse now dominated the building (figure 24). While the new 500,000 gallons-per-day (1893 kiloliters-per-day) Cameron pump was being installed, only the old water-powered pump remained operational. To meet the demand, Water Superintendent Erwin Muschlitz pressed the steam pumping engine of the Reliance Fire Company into service.[55] The portable steamer ran continuously for ten weeks before its boiler finally gave out. Fortunately the new pump was almost ready; and on July 3rd the Bethlehem "Daily Times" jubilantly announced, "The steam pump is in operation."[56]

By the early 1880's the town again needed more water than the waterworks could deliver. A second steam pump having a capacity of 500,000 gallons (1893 kiloliters) per day was purchased from the Worthington Company in 1885

24. This view of the oil mill building from the southwest shows the 1874 boilerhouse addition on the right with a 45 foot (12.8 m) high chimney at the south end.[54] *Since the 1889 iron chimney (figures 23 & 26) is absent, this photograph must have been taken between 1874 and 1889.*

25. This 1874 watercolor drawing by Rufus Grider shows the north side of the oil mill.[57] *The entrance to the residence of the waterworks engineer is on the right. The headrace flows from the Monocacy Creek at the right edge of the picture to the wheelpit behind the bottom center window. Water Street makes a sharp bend as it passes between the 1762 waterworks and the east end of the oil mill; this was considered a traffic hazard in the 1930's and was a factor in the demolition of the mill.*[58]

26. A 60 foot (18 m) tall iron stack was added on the south side of the old stone building in 1889 to service the two new boilers.[66] The old dam (left) was no longer a source of power, and the tailrace seen in figure 24 had been filled in.

for $1790 and installed at the oil mill.[59] Then in 1888 a new problem appeared. Water was being drawn from the spring faster than the surrounding limestone could purify and renew it; pollution from the nearby homes, creek, and millrace was inevitable.[60] The Town Council ordered the spring deepened by blasting in hopes of obtaining more pure water. The experiment was successful, and the flow was more than doubled to two million gallons (7570 kiloliters) daily.[61]

By 1888 the Bethlehem waterworks was again operating at maximum pumping capacity, and further expansion was necessary.[62] Following the recommendations of engineer C. W. Knight,[63] a new boiler system and a new larger pump were installed to gain maximum fuel economy. This necessitated a complete renovation of the oil mill building. The waterwheel, the 1832 water-powered pump, the grist milling machinery, the residence for the waterworks engineer,[64] and everything else "from the ground to the upper joists" was removed. The central wheelpit was filled in to provide a foundation for the two new boilers (figure 23), and a 60 foot (18 m) iron chimney was erected on the south side of the old stone building (figure 26).[65] The level of the ground floor was raised 3 feet (0.9 m) to protect the machinery from floods. "One noncondensing compound duplex pumping engine of the

Deane Steam Pump Company, having a capacity to deliver two million gallons (7570 kiloliters) in twenty-four hours" was purchased.[67] It was installed in the eastern part of the building in October 1889, and the old Worthington and Cameron pumps were retained as auxiliaries (figure 23). The western part of the building was converted to coal storage for the boilers.

Early in the twentieth century water quality became a serious problem. The limestone stratum in which the spring was located[68] simply could not filter out all the wastes generated by the several thousand people living directly uphill from the spring. Epidemics of typhoid and dysentery occurred and were traced to the water supply.[69] By 1911 the town was forced to install a germicidal plant to process the once-pure water.[70] Seeking a new source of supply, the borough sank wells at Illick's Mill, 2.5 miles (4 km) up the Monocacy valley from the oil mill, and a new waterworks was constructed there.[71] An electrically-powered De Laval centrifugal pump purchased in 1911 to increase pumping capacity at the oil mill waterworks was installed at the Illick's Mill site instead.[72] Under orders of the Pennsylvania Commissioner of Health the steam pumps at the oil mill were shut down permanently in January 1913.[73] The spring was sealed, all connections with the town water system were "absolutely severed," and the building which had served Bethlehem for so long was left to the pigeons and the rats.

FINAL YEARS 1913-1934

After the pumps were shut down, the old oil mill building stood vacant for twenty-one years. It deteriorated from lack of maintenance and became an eyesore and disgrace to the community. Many local citizens wanted to restore it for some public use or simply preserve it for its historical significance,[74] but Bethlehem's elected leaders decreed a different fate for the old stone structure.

In November 1933, as part of the program to overcome the Great Depression, President Franklin Roosevelt established the Civil Works Administration. It provided nationally-funded jobs for local projects requiring little or no planning or material.[75] Northampton County's program was supervised by Robert L. Fox, Bethlehem's City Engineer. He initially approved six projects within the city, among which was the only demolition project in the county: "Razing the old water works building on Water Works Road (two story stone building, 30 by 40), $1760."[76] The proposal generated a storm of protest.[74] In response Mayor Robert Pfeifle claimed he was "endeavoring to obtain funds from the State for the repairing of the building in order to preserve it."[77] However, the funds never materialized.

Razing was still "under consideration" on December 9, 1933.[78] Then the building vanished, both from the site and from the records.[79] Local oral history sources state

27. The Bethlehem oil mill was torn down in 1934, and only its grass-covered foundation remains today. The outline shape of the final building (dashed lines) can still be seen at the edge of Monocacy Creek (lower left) near the restored 1762 waterworks (upper right).[82]

that it was demolished by the WPA[80] during the 1930's, presumably in early 1934.

Even the foundation disappeared beneath weeds and brush. In 1968 an archeological team located and marked the remains but did not excavate them.[81] Today only the grass-covered outline of the foundation and a sign mark the location of Christensen's remarkable oil mill (figure 27) which played such an important part in the life of Bethlehem for one and a half centuries.

TECHNOLOGY OF THE OIL MILL INDUSTRIES - I

The wealth of surviving documentation on the 1765 Bethlehem oil mill fortunately includes many details of the machinery and the processes used inside the building. This information provides some important insights into both the practices and the origins of early technology in Pennsylvania. One can clearly see how eighteenth century Germanic milling techniques for pressing oilseeds, softening hemp, stamping tanbark, and grinding grains were brought to Pennsylvania by the Moravian colonists who settled at Bethlehem.

In this chapter and the one that follows, we present detailed reconstructions of the five manufacturing processes carried out at the 1765 Bethlehem oil mill during the 1766-1800 period. Since this information might someday prove useful for reconstruction of the 1765 oil mill with all its machinery, both rigorously documented and tentative conclusions on these processes are included. We also discuss in detail the water pumping technology employed in the building from 1832 to 1913 when it served as the town waterworks.

1766 HERRNHUT DRAWINGS

The most important source of information on the technology of the early oil mill industries is the remarkable set of contemporary machinery drawings reproduced in figures 28-31. These give both plan and elevation views of the two waterwheels, the oilseed stamping and pressing machinery, the two sets of groat millstones, the tanbark/snuff stamping equipment, and the hemp stamping mill in the 1765 oil mill building. Comparison of these drawings with eighteenth century German milling texts can give us an accurate interpretation of how this specific machinery was used.

These plans are in the collection of the Moravian Archives at Herrnhut, East Germany and were first published in 1929 by Erbe.[4] Although undated, the drawings undoubtedly were made in Bethlehem in 1766 to report back to the Moravian Church headquarters at Herrnhut concerning the newly-constructed 1765 oil mill. A 1766 entry in the "Oil Mill Account Book" reads:

To Brother Schulze for
2 plans of the oil mill
1 plan of the waterworks
Made for David Nitschmann £ 1.7.6 [5]

The machinery shown in these drawings fully matches the machinery descriptions in the 1765-1766 portions of the "Oil Mill Diary" and the "Oil Mill Account Book." The absence of the hemp roller mill from these plans also points to a 1766 date, since the hemp rolling machinery was not installed until 1767-1768.[2]

WATERPOWER SYSTEM

The waterpower for the 1765 oil mill came from the adjacent Monocacy Creek. A low dam just northwest of the building (figure 56) diverted water to the oil mill as well as to the 1762 waterworks located 30 feet (9 m) further east. The precise construction of this dam is unknown. It apparently had a wooden frame which was built in sections on shore and then installed in the creek, probably by loading with stones.[6]

Twin undershot waterwheels in the center of the building furnished power to the various mills. This side-by-side location of the wheels minimized the excavation needed for the wheelpit and raceway, while enclosing the wheels inside the building helped to reduce icing during the winter. The two waterwheels used water from a common headrace, though each had its own sluice gate. Separate wheels allowed operational flexibility, since two mills could be run simultaneously when water was plentiful. On the other hand, one wheel was powerful enough to operate one mill during periods of low demand without wasting water.

Each wheel was 17.5 feet (5.33 m) in diameter and 4.5 feet (1.37 m) wide (figure 29). The available net head of water was about 2 feet (0.6 m).[7] It is impossible to determine a reliable power rating for these wheels, for many details of the sluice, wheelpit, and wheel design are unknown. However, if one assumes that the hydraulic system and machinery were well-designed and constructed like those of the 1762 waterworks, each wheel might have rotated at 7.5 rpm and have developed 5 to 6 horsepower (3.7 to 4.5 kw),[8] of which at least half was immediately lost to friction and hysteresis before any useful work was done.[9]

In retrospect, it is apparent that the power requirements of the 1765 oil mill exceeded the waterpower resources available at the site. The Monocacy Creek was used to drive three waterwheels (two at the oil mill, one at the adjacent waterworks) from one low dam without a sizable storage reservoir. As Bethlehem grew, there were times when the oil mill had to shut down so that the waterworks could pump sufficient water to the town.

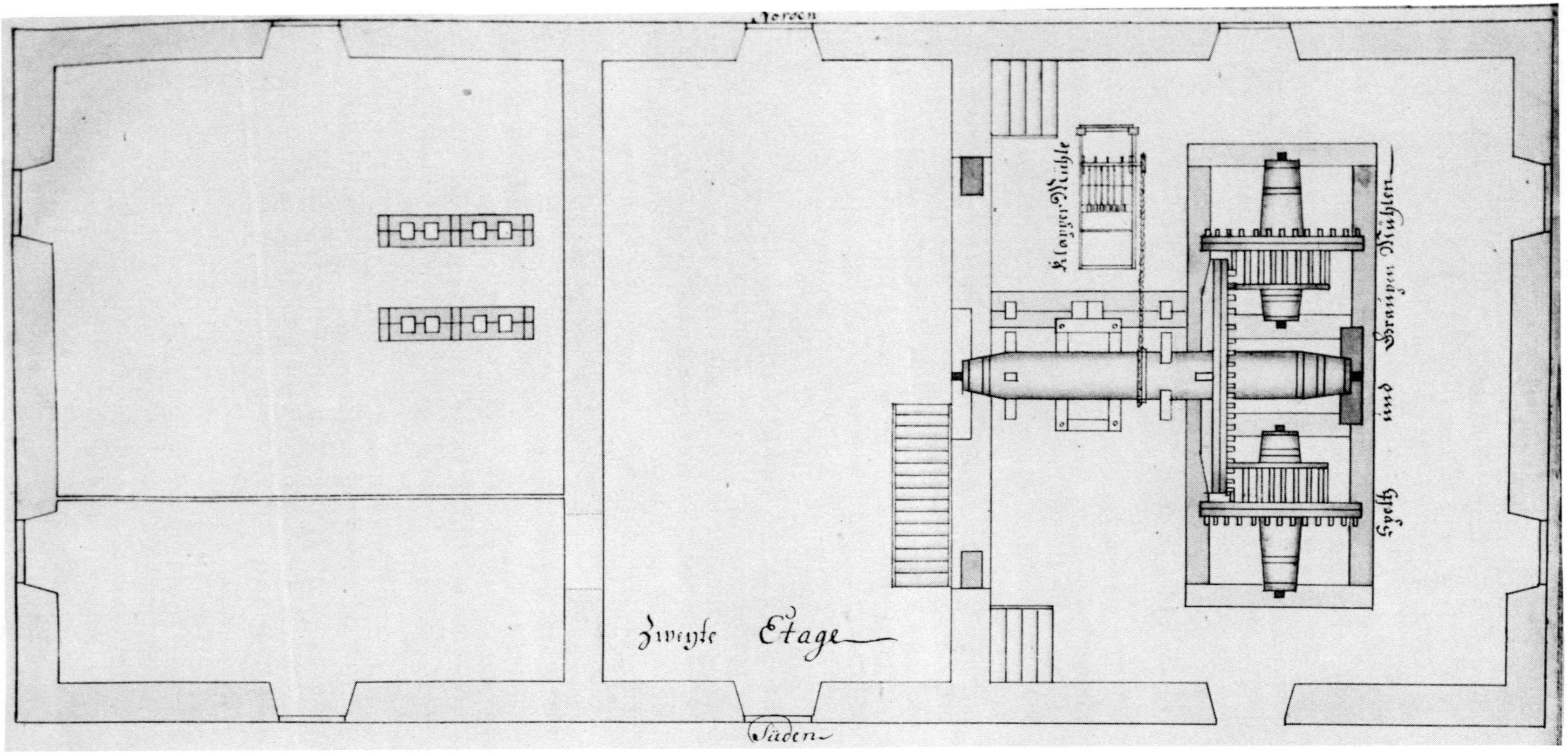

28. *Plan of the second floor of the 1765 Bethlehem oil mill showing the gearing for the groat mill (right) and the adjacent clapper mill.*[1] *The central area was a raised floor above the two waterwheels. The hemp roller mill was installed in the western room a year after this drawing was made.*[2] *Scale same as figure 29.*

29. *Plan drawing of the machinery on the first floor of the 1765 Bethlehem oil mill showing the hemp stamping mill (upper left), the tanbark mill (lower left), the waterwheels (center), and the oilseed mill (right).*[1] *This plan was drawn by Schulze in 1766 when construction of the new mill was practically complete. Scale (bottom) divided into one and ten foot (0.3 and 3.0 m) lengths.*

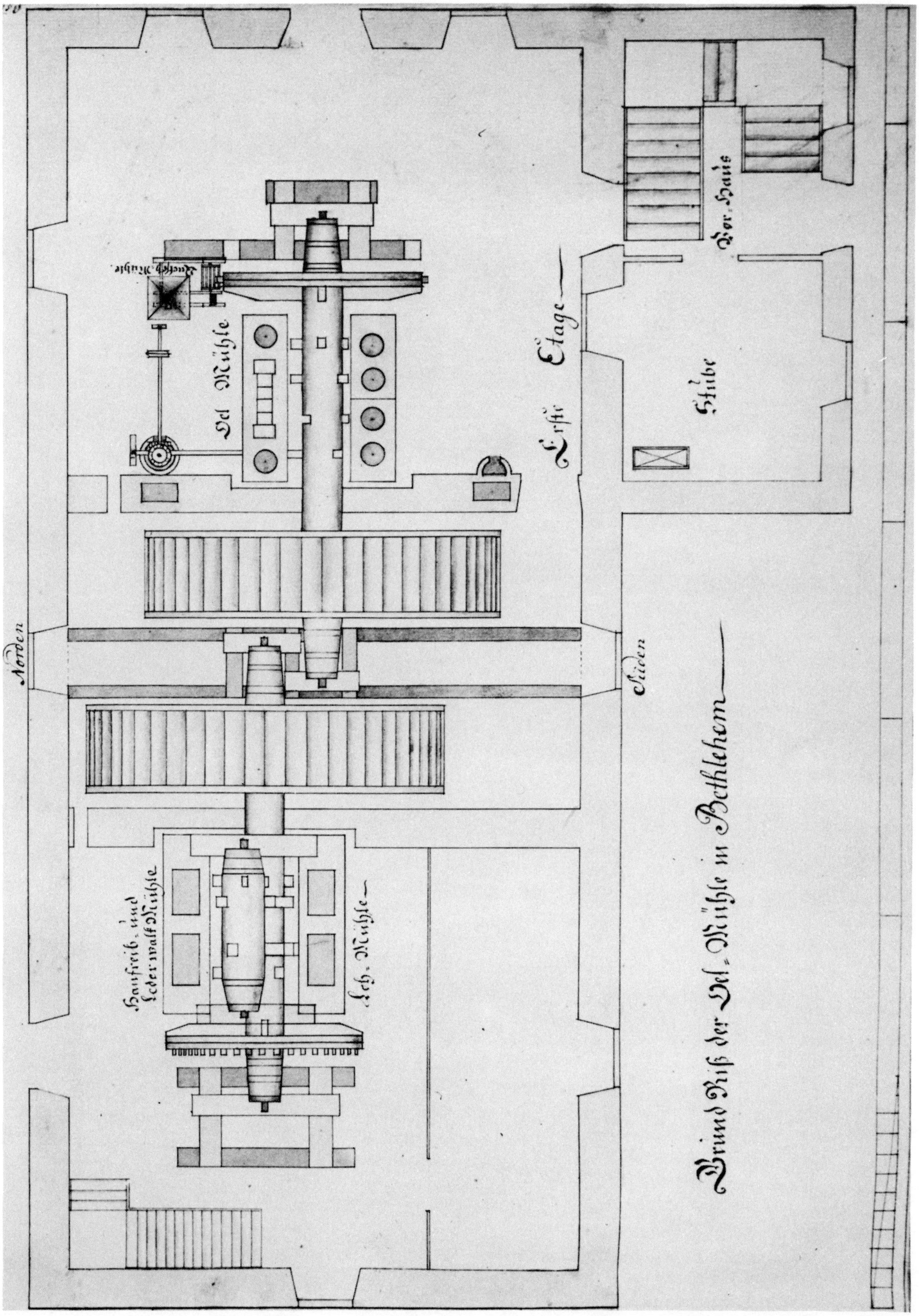
Grund Riß der Oel=Mühle in Bethlehem
Erste Etage
Oel Mühle
Hanfreib. und Lederwalk Mühle
Loh. Mühle
Norden
Süden
Stube
Vor=Haus

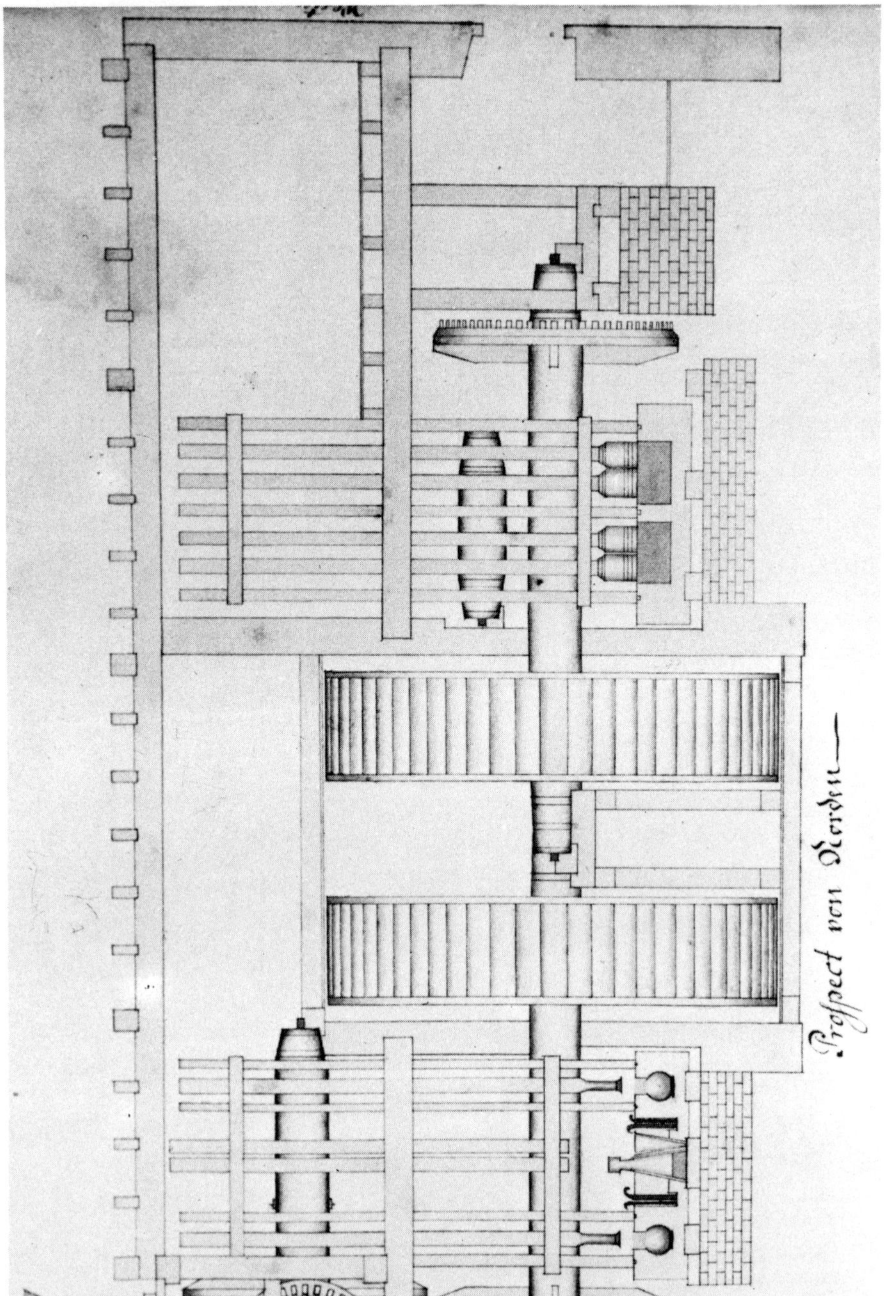

30. North elevation of the machinery in the 1765 Bethlehem oil mill showing the groat mill (upper left), the oilseed press and stampers (left), the two waterwheels (center), and the hemp stampers (right).[3] Scale same as figure 31.

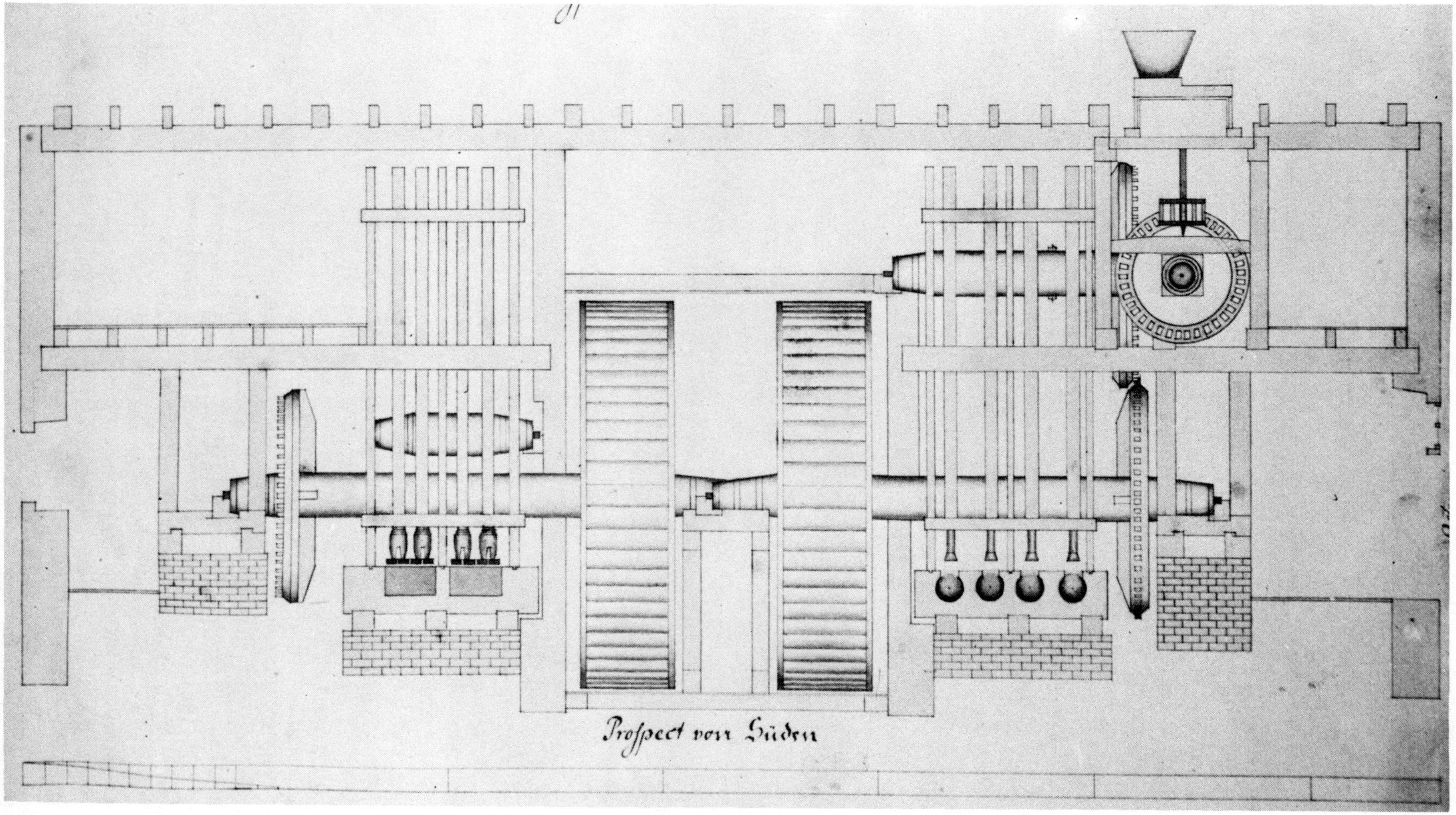

31. South elevation of the machinery in the 1765 Bethlehem oil mill showing the tanbark stampers (left), the two waterwheels (center), the oilseed stampers (right), and the groat mill (upper right).[3] *Note that the large face gear at the end of the west waterwheel shaft served no useful purpose in 1766 when Schulze made this drawing. It was intended to drive the hemp roller mill installed on the floor above in 1767-1768.*[2] *Scale (bottom) divided into one and ten foot (0.3 and 3.0 m) lengths.*

OILSEED MILL

The machinery for the oilseed mill is shown in considerable detail in the 1766 Herrnhut drawings (figures 28-31). As one would expect, this machinery resembles the oilseed crushing and pressing equipment found in central Europe during the late eighteenth century. Using the surviving records from Bethlehem, descriptions of similar European oilseed mills,[10] and a contemporary report about how a colonial Pennsylvania oil mill was operated,[11] we can readily deduce how linseed oil was produced from flaxseed at Bethlehem (figure 32).

Flaxseed was a by-product of the flax fiber crop and was received at the oil mill in bags or in baskets. If the seed was too damp, so that it might become moldy during storage, it probably was spread out on the floor of the loft and dried for a few days.[12] Before crushing the seed was cleaned to get rid of both the fine (dust, dirt, broken seeds) and the coarse impurities (sticks, stones, leaves).[13] The sieving machine in the groat mill (the clapper mill depicted in figure 44) was used for this purpose. However, the fanning machine (figure 43) probably also was used to clean flaxseed.[14]

The next step was to crush the flaxseed in the four stamping mortars on the southern side of the oil mill (figures 29 & 31).[15] Breaking up the seed greatly aided oil release during the subsequent pressing. The construction and operation of such a stamping mortar is shown in figure 33. The mortar, 17 inches (43 cm) deep, was hollowed out of a large base log, and an iron plate was installed on its bottom surface.[17] The heavy wooden stamping pole was 7 inches (18 cm) square and 18 feet (5.5 m) high with an iron ring around the bottom tip and a short side arm for lifting. The four stampers were lifted in rotation by a revolving cam shaft and then released to fall into the mortar bowls. While filling or emptying a mortar, the operator raised the stamping pole out of action by means of a rope, lever, and peg mechanism. Flaxseed was thoroughly crushed to an oily meal in these mortars and then taken to the roasting pan.

The third step in producing linseed oil was to roast the flaxseed meal. Brief heating before the pressing increased oil yields by causing further rupture of the oil-containing cells in the seed. Sometimes the roasting step was omitted, however; and a lower yield of lighter colored, better quality "cold pressed" oil was produced.[18] At contemporary oil mills in Europe,[10] the roasting process was carried out in a large round iron pan heated on a stove. Constant mechanical stirring was always employed to avoid charring the seed with consequent oil loss.

The 1766 plan drawing of the ground floor of the Bethlehem oil mill (figure 29) shows a circular device (just west of the *Quetsch-Mühle*) that appears to be such a seed roaster. This device is round (the roasting pan[20]), is located next to a wall with a chimney flue[21] (for the stove heating the pan), and contains gearing and

32. Linseed oil was produced from flaxseed at the Bethlehem oil mill using a four-step process. Recycling the press cake for a second run yielded additional oil.

33. Probable cross section of the stamping mortars used to crush flaxseed and press cake.[16] The rotating cam lifted the stamper by means of the side arm and then released it to fall upon the seed in the mortar bowl below. When filling or emptying the mortar, the stamper was disengaged from the cam and held in the up position by raising a lever against a peg on the side of the stamper.

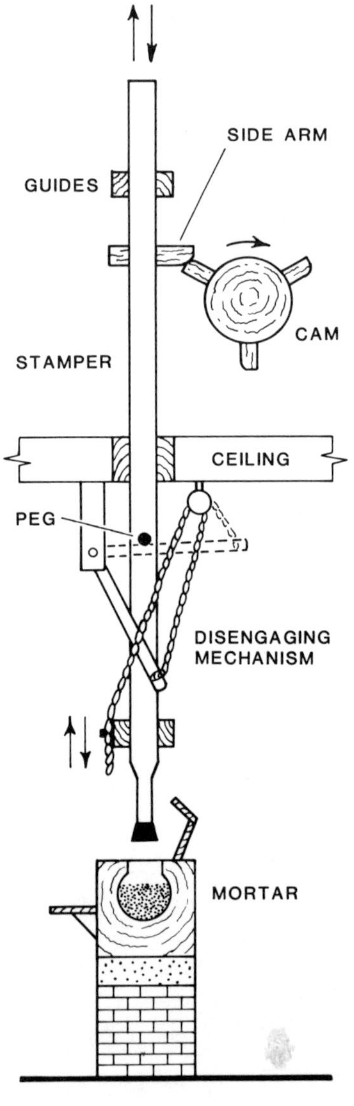

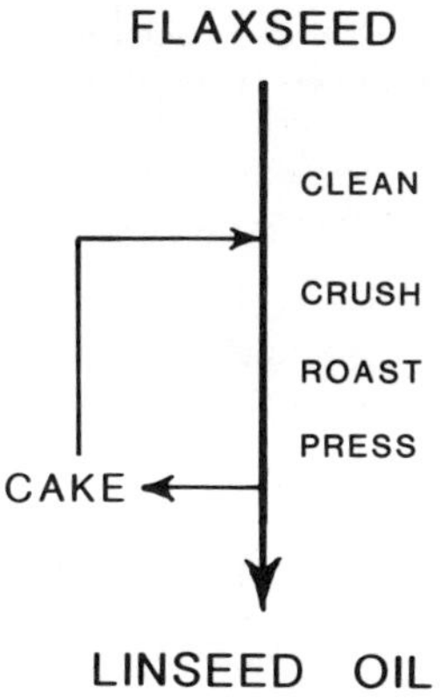

shafting (the power drive for the stirrer[22]). A possible reconstruction of this seed roaster is presented in figure 34. After 5 to 10 minutes of roasting in this apparatus, the flaxseed meal was placed in woolen bags and then taken to the adjacent wedge press.[23]

The final step in the process was to press the linseed oil from the crushed and roasted flaxseed using a wedge press (figure 35). This press consisted of a large wooden beam (base log) 27 inches (69 cm) square with a rectangular trough hollowed out of the center. The bags of roasted seed were wrapped in leather-backed horsehair mats (the "hairs,"[24] figure 62) and then placed between heated iron plates at each end of the trough. The remaining space between them was filled with two wooden wedges and several spacer blocks. One of the wedges was pointed downward so that blows on its top surface would produce lateral pressure to express the oil. The other was

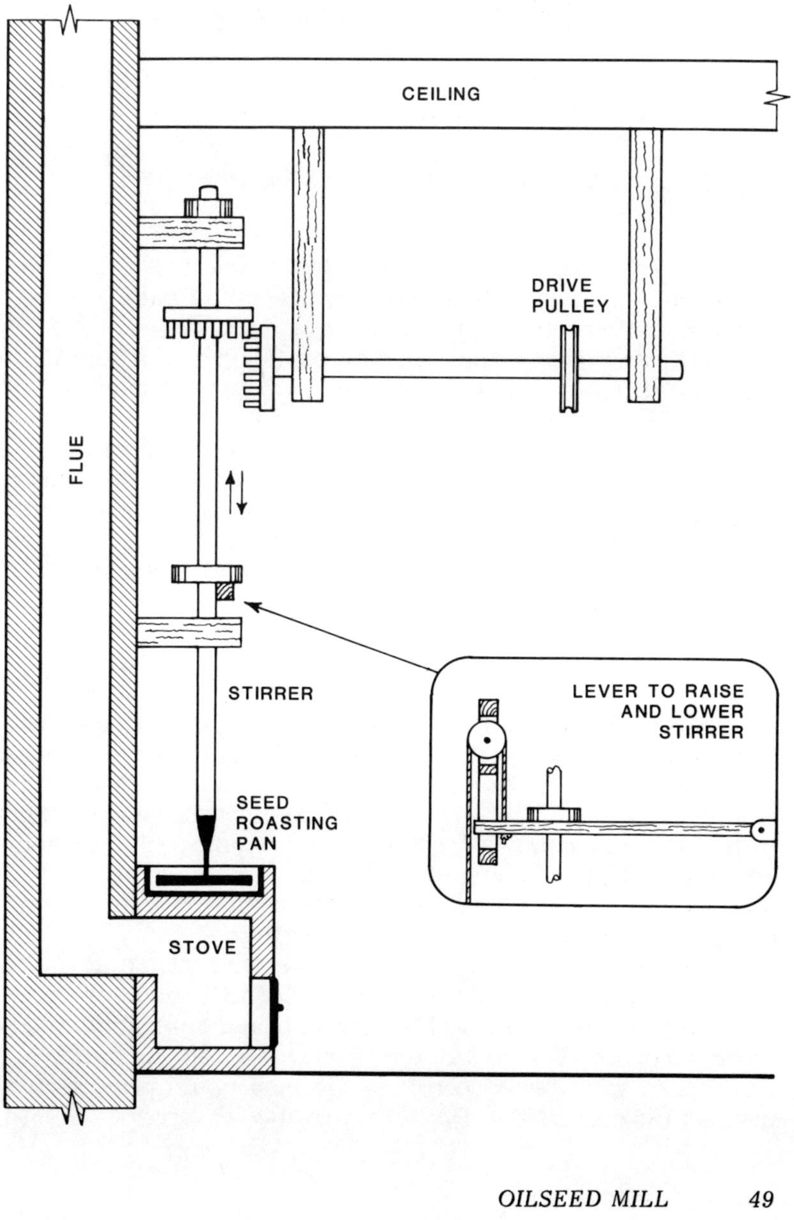

34. The apparatus for roasting crushed flaxseed at the Bethlehem oil mill probably was very similar to this conjectural reconstruction derived from details pictured in figure 29.[19] *To increase the yield of oil, crushed flaxseed was placed in an iron pan and heated on a stove for several minutes just prior to pressing. An inverted-T stirrer powered from above prevented the seed meal from charring during heating.*

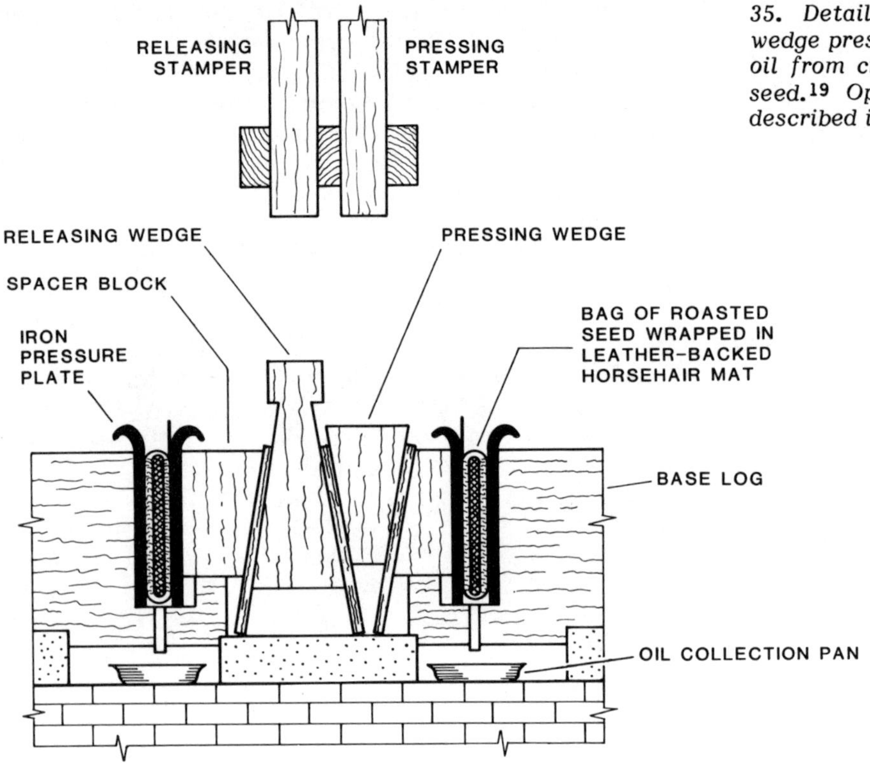

35. Detailed cross section of the wedge press used to express linseed oil from crushed and roasted flaxseed.[19] *Operation of the press is described in the text.*

an inverted wedge that would release the press components when it was hammered downward. Heavy wooden stamping poles were positioned above each wedge.

With the press thus assembled (figure 35), the idle pressing stamper was engaged with its lifting cam (as in figure 33) so that it was repeatedly lifted and dropped upon the pressing wedge. Some forty to sixty blows on that wedge compressed the bags of roasted seed enough to cause linseed oil to trickle downward and drain into collector pans beneath the base log. After this oil flow stopped, the releasing wedge was pounded several times with the releasing stamper; and the press could then be disassembled and another run started. The oil produced was then poured into wooden kegs for storage and shipping.[25]

After disassembling the press, the woolen bags were peeled off the hard, brittle press cake; and the cake was coarsely ground in the adjacent crushing mill (the *Quetsch-Mühle* in figure 29). This ground cake was reprocessed in the same manner as fresh flaxseed (stamped,[15] roasted, and pressed) to recover an additional amount of linseed oil. The cake from this second pressing was then ground and used for either cattle feed or fertilizer.

Flaxseed processing was mainly a wintertime activity at the mill. The crushing started in late October or early November soon after the flax harvest and continued more or less regularly through the winter and spring until all the seed was processed.[26] Two men could process ten bushels (352 liters) of flaxseed per day to produce 14 to

17.5 gallons (53 to 66 liters) of linseed oil.[27] Highest yields were obtained when less seed was put into the press bags and the bags were left in the press longer. The sound of the stamping poles falling every few seconds made the oilseed mill a noisy place. Anyone working there for a number of years probably suffered from "oil miller's disease" (partial deafness) as a result.

Linseed oil was not the only oil produced at the Bethlehem oil mill. Hempseed, a by-product of the hemp crop grown for fiber, was crushed in the same manner as flaxseed to produce hempseed oil,[28] a product that could partially substitute for linseed oil in paints. In 1768 Dr. John Matthew Otto, a Bethlehem physician and apothecary, experimented with making sunflower seed oil and cottonseed oil at the oilseed mill.[29] Both types of seeds yielded 3 quarts (2.8 liters) of oil per bushel (35.2 liters) of seed. The sunflower oil was "frequently used by our brethren for salad," while the cottonseed oil served as a medicine for the treatment of colic. John Arbo also experimented with extracting fatty oils from unusual seeds at the Bethlehem oil mill (figure 13).

HEMP MILL

Following the tradition of their Germanic homelands, the Moravians utilized hemp as a major fiber for making strong, durable cloth. Hemp fiber was spun into hempen yarn and then woven into hempen cloth.[30] To make the coarse hemp fibers soft and pliable enough for normal spinning and weaving, it was necessary to hammer, stamp, or roll them to break down their natural resinous coating and to subdivide the fibers. This softening process was accomplished using machinery at the Bethlehem oil mill.

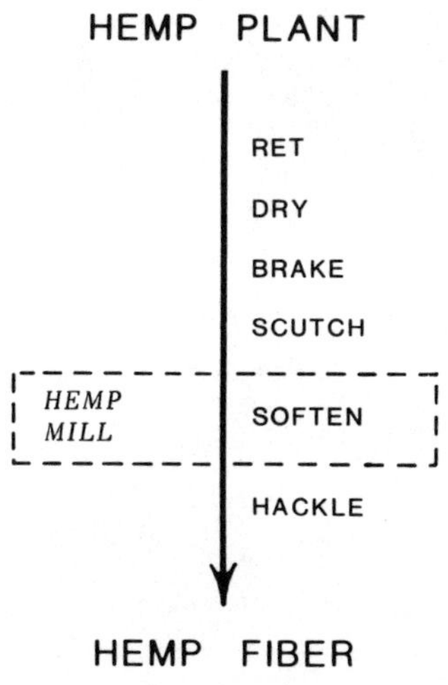

36. The Moravian process for preparing hemp fiber soft enough for spinning and weaving into cloth required six steps. Only the softening step (stamping or rolling the coarse fibers to split them into finer fibers) was done at the oil mill.

The Moravian process for preparing spinnable fiber from the hemp plant required six steps (figure 36) and closely resembled the preparation of linen fiber from flax.[31] Hemp fibers occur only in the outer layer of the stem of the plant and must first be separated from the other plant tissues. To accomplish this, the freshly pulled hemp stalks were "retted," i.e., laid on the ground and exposed to the weather for 3 to 6 weeks until the nonfibrous portion of the stem started to rot and decompose.[32] Next the hemp stalks were dried by heating them over a fire.[33] The dry stalks were then crushed using a hemp brake, which resembled a common flax brake but was larger. This shattered the friable, rotted portion of the stalk so that most of it fell away from the fibers. The remaining bits of woody material were removed by scutching, i.e., by scraping the fibers with a wooden knife. To soften the fibers for spinning, they were repeatedly hammered, stamped, or rolled with a stone. Finally, the fibers were pulled through a series of hackles, combs with iron teeth, to align the fibers and to remove the shorter lengths.[34]

Although spinnable hemp could be produced by hand labor alone, softening the fibers by pounding them with a

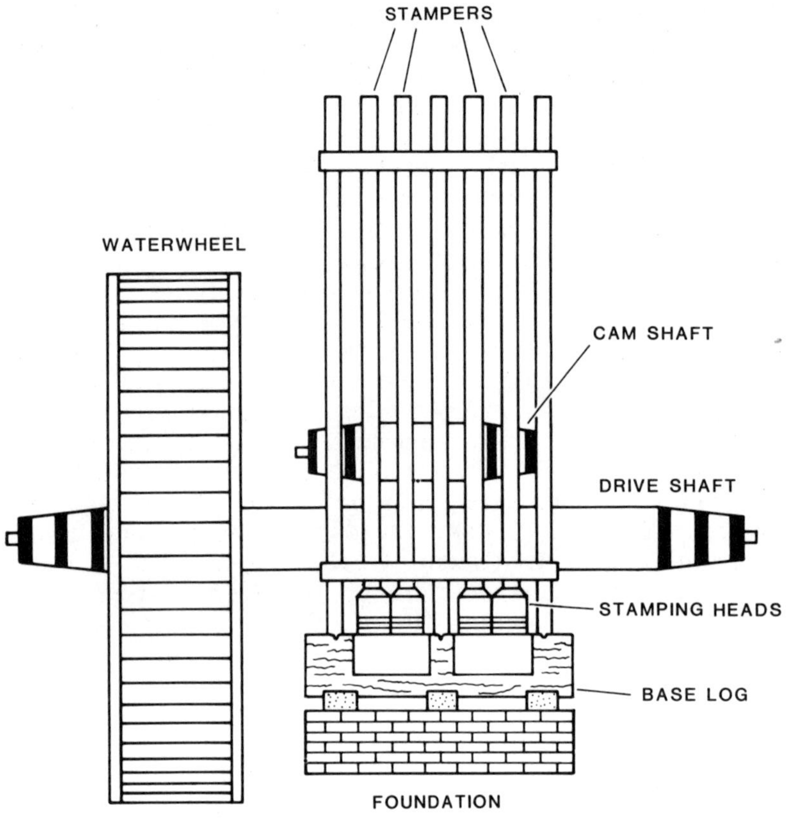

37. Construction of the hemp stamping machinery installed at the Bethlehem oil mill in 1765 (redrawn from figure 30 with the base log repositioned on the foundation). This equipment was built for both stamping hemp and fulling leather, but it was not used to any extent for the latter process.

hammer was extremely arduous work. At Bethlehem this softening step was mechanized using water-powered equipment of traditional Germanic design.[35] Hemp stamping mills were built in the 1752 and the 1765 oil mill buildings, and in 1767-1768 a hemp roller mill was added. Many mechanical details of the two machines at the 1765 oil mill have survived so that we can clearly understand how they operated.

The hemp stamping equipment at the 1765 Bethlehem oil mill is depicted in the 1766 Herrnhut drawings (figures 29 & 30); a simplified elevation view has been drawn in figure 37. It was a stamping mill very similar in size and design to the oilseed mill installation, except for the different shapes of the stamping heads and trough. Hemp fiber (after scutching) was placed in the stamping trough, the stampers were engaged with the lifting cams[36] as shown in figure 33, and the heavy stamping poles were repeatedly lifted and allowed to fall onto the hemp. After the hemp had been sufficiently softened, the stampers were disengaged, the hemp was removed from the trough, and it was ready for the final hackling step.

The stamping heads shown in the 1766 Herrnhut drawings (figure 30) appear to be stepped like those used in fulling mills.[37] Although there is no confirming side view, this interpretation seems likely for two other reasons. This was a dual purpose mill designed to full leather as well as to stamp hemp. Moreover, before the 1752

38. The heads on the hemp stampers in the 1766 Herrnhut drawings (figure 30) resemble those of a fulling mill. A conjectural cross section of the head and trough design is shown here. The stepped head served to rotate the fiber with each blow because of the curved side and bottom of the trough.

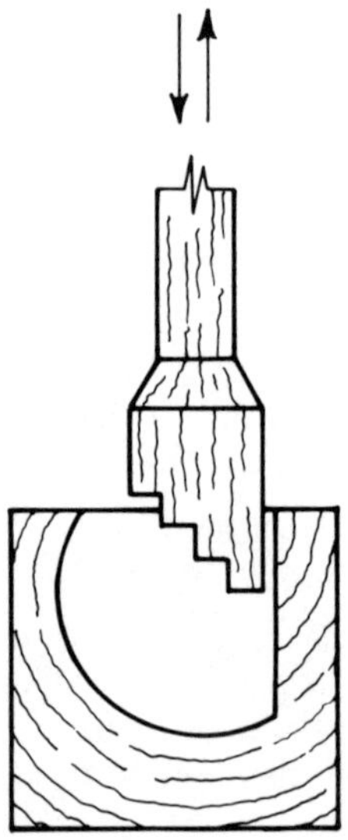

39. The new hemp rolling equipment installed at the Bethlehem oil mill in 1767-1768 closely resembled this hemp mill in a German museum.[44] Bundles of hemp fiber placed on the circular bed were softened by rolling the heavy stone over them for a period of time.

hemp stamping mill was built, the Moravians had used their cloth fulling mill to stamp hemp.[38] A conjectural cross section of the hemp stamping head and trough (figure 38) shows how the stepped head could have rotated the fiber with each blow because of the curved sides of the trough, the same as in a fulling mill.[39]

In 1767-1768 another type of hemp softening machinery was installed at the oil mill.[2] This was a hemp roller mill located on the second floor of the west side of the building in the vacant area shown in the 1766 Herrnhut drawings (figure 28).[40] The description of materials purchased for constructing this new hemp mill strongly suggests that it consisted of a stone that rolled around a circular bed.[41] Similar hemp mills have been reported in early America,[42] and one has survived intact at a German museum (figure 39).[43]

To soften hemp on such a roller mill, bundles of scutched fiber were placed on a circular bed, and a heavy stone shaped like a truncated cone (figure 63) was rolled over the fibers until they were sufficiently soft and pliable. An iron axle through the center of the stone was attached to a vertical drive shaft at the center of the bed. At the Bethlehem installation the hemp roller mill was probably driven from below by a gear train that included the large face gear at the end of the west waterwheel shaft (figure 31). An iron collar around the edge of the bed kept the fiber in the pathway of the revolving stone. From time to time during the run, the hemp was turned and shaken to assure that all parts were evenly treated. At the end of the run, the roller mill was stopped, and the softened hemp was removed and taken for the final hackling. Similar hemp mills in Europe reportedly could soften two hundred pounds (91 kg) of hemp fiber in eight hours.[45]

The fact that the Moravians built a second type of hemp softening mill in 1767-1768, only two years after completing the 1765 machinery, implies that their 1765 hemp stamping mill did not yield satisfactory results. In their 1765 advertisement (figure 15) they had claimed a distinct advantage for their stamping mill: "The Hemp is not rubbed[46] with a Pumicestone [roller stone] in the common Way, that being attended with many Dangers; but it is stamped in a particular Manner and becomes pliabler and fitter than with the Stone." Apparently this claim was not realized. Their customers may have been quite satisfied with the "common" hemp roller mill, or perhaps the roller mill processed hemp with less manpower and thus at less cost.[47]

Hemp usually was processed at the mill during the late fall and winter months. In 1766-1767 when only the stamping mill was used, 6242 pounds (2328 kg) of hemp were softened. With both hemp mills in operation, the Bethlehem oil mill processed 28,502 pounds (10,631 kg) of hemp in 1770-1771 and 41,403 pounds (15,443 kg) in 1771-1772.[48] Hemp softening continued at least through 1792.[49] No mention of the process after that date has been found.

TANBARK MILL

About 300 feet (91 m) north of the oil mill the Moravians operated a large tannery. "It is one of our most profitable and most essential trades," wrote Bishop Cammerhoff.[50] "[The tannery] supplies all our shoemak-

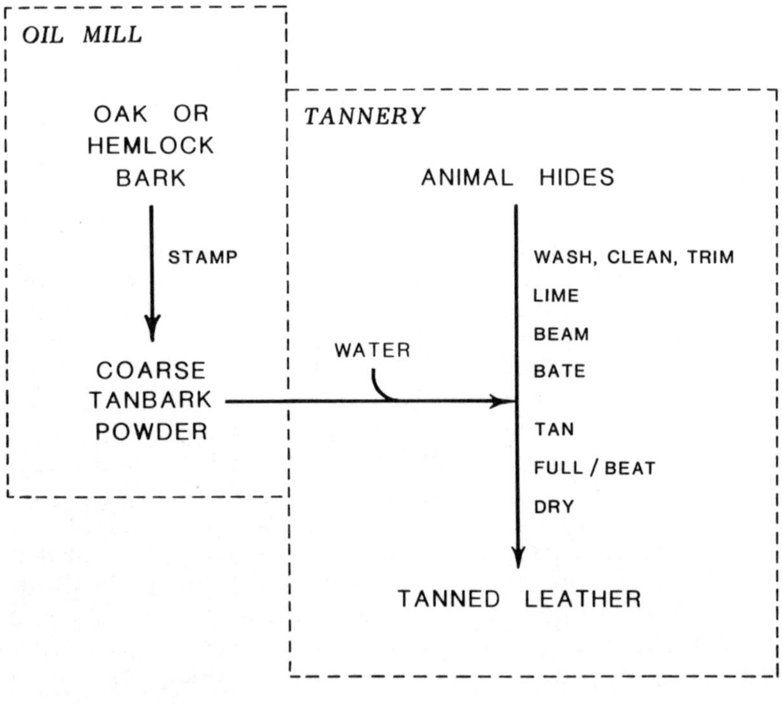

40. The tanbark mill was used to stamp oak and hemlock bark into a very coarse powder for use at the Bethlehem tannery. The tanner used this product in his tanning vats where the water leached the tannins out of the bark for chemical reaction with the animal skins. Tanning was only one of seven steps used to prepare tanned leather from animal hides.[52]

41. Construction of the tanbark stamping machinery installed at the Bethlehem oil mill in 1766 (redrawn from figure 31). The knives on the bottoms of the stamping heads reduced oak or hemlock bark to a very coarse powder.

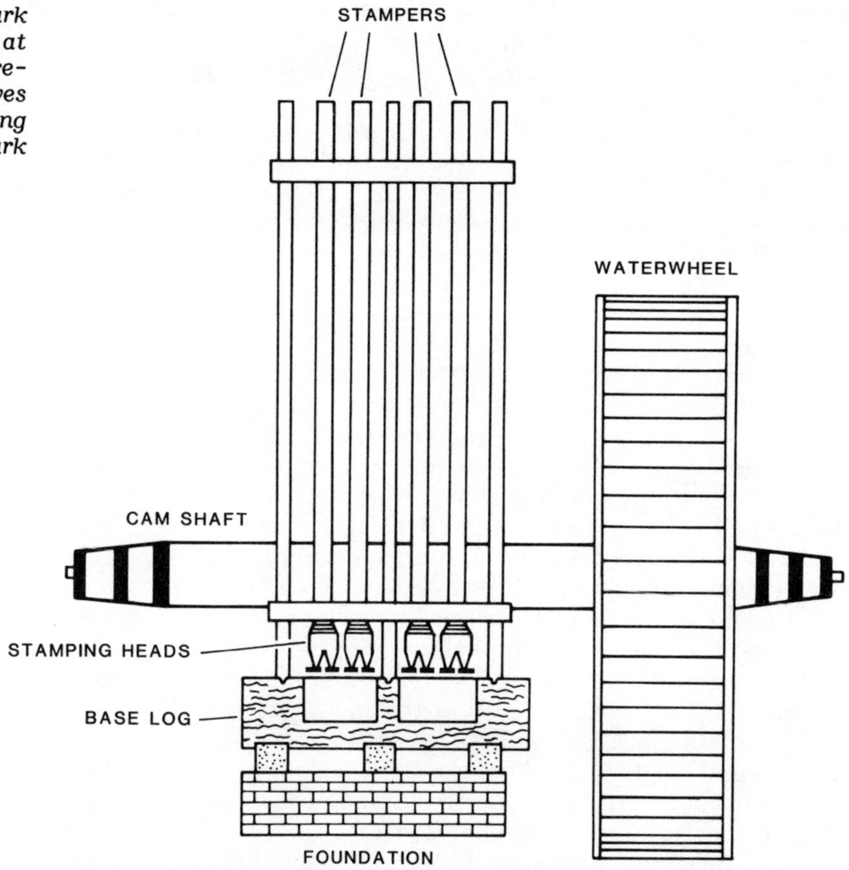

ing trades . . . and all our people with leather." From 1752 to 1807 the tanbark used by this tannery was stamped on tanbark mills located in the 1752 and 1765 oil mill buildings.[51]

The conversion of animal hides into tanned leather was a lengthy, seven-step process in the late eighteenth century (figure 40). First the hides had to be washed, cleaned, and trimmed. Next they were soaked in a lime solution for several weeks; this facilitated "beaming," the subsequent scraping of the hide to remove the hair and any remaining flesh. The hides were then soaked for many weeks in a "bating" solution, a foul-smelling mixture of water, salt, and manure that neutralized any remaining lime and also altered the chemical structure of the skin to make it more flexible. Next came the actual tanning step of immersing the hides in a vat filled with tanbark and its rich "liquor" extract for six to twelve months. Afterwards the leather was beaten or fulled to make it more pliable and then carefully dried. Soft, waterproof leathers with a smooth finish required further treatment by a currier.[53]

The principal step in the transformation from hide to leather was, of course, accomplished by the long soak in the tanning vats. Here astringent phenolic substances known as tannins, leached from a suitable tree bark, reacted chemically with the animal skins to produce a durable leather. Each hide required one to two times its weight in bark for proper tanning.[54] Oak and hemlock barks were most commonly used; so were oak galls, a tumorous bark growth.[55] Tanbark was usually gathered in the spring, when it was easiest to peel the logs, and then thoroughly dried. Records show that Bethlehem's tannery consumed about 60 to 100 cords (218 to 363 cubic meters) of tanbark annually.[56] The bark was stored in a shed just north of the oil mill (figure 3) to prevent rainwater from washing out the valuable tannins. Tanbark was stamped as needed.

Before the bark could be used for tanning, it was reduced to a very coarse powder with a tanbark mill. A coarse product was desirable for several reasons: small particles could be handled more easily than a fine powder, water would penetrate better for leaching out the tannins, the tannin-rich liquor would filter out more readily, and the tan vat residue would dry more quickly.[57] Most early American bark mills were animal-powered and consisted of a large cylindrical stone rolling around a circular bed, crushing the bark with its weight.[58] The Moravians, however, followed contemporary German practice[59] and built a bark stamping mill in Bethlehem.

The tanbark stamping machinery installed at the Bethlehem oil mill in 1766 is shown in figure 41. Four stamping poles 18 feet (5.5 m) high fell upon bark placed in two mortars hollowed out of a large base log c. 9 feet (2.7 m) long and 26 inches (66 cm) square. Each stamper had a large iron head with cutting knives on the bottom.[60] Lifter cams on the waterwheel shaft raised the stampers (as in figure 33) about 2 feet (61 cm), then released them to fall and chop up the bark in the troughs. The vibrations from the alternately falling stampers and the asymmetrical shape of the mortar bowls continually shifted the bark for even cutting. To contain the inevitable dust resulting from stamping dry tanbark, a wooden case was built around the mortars shortly after the mill was installed.[61] From 17 to 23 mortarloads of tanbark could be stamped in one day, so the loading/stamping/unloading cycle must have lasted about one hour. The production capacity of the mill was c.175 cords (635 cubic meters) of raw bark per year.[62]

The 1766 tanbark mill remained in use for four decades. The records show a major repair or improvement of the machinery in 1794, but no details of this alteration have been found.[63] In 1806 the tanner complained that the continual shortage of water prevented the oil mill from stamping as much tanbark as he needed.[64] A new horse-driven tanbark mill was installed in a separate building, and tanbark stamping at the Bethlehem oil mill ended in 1807.[65]

TECHNOLOGY OF THE OIL MILL INDUSTRIES - II

GROAT MILL

Groats, coarsely ground cereal grains, were a regular part of the diet for Bethlehem's early population. They were produced from barley, oats, buckwheat, and millet by first removing the outer hull of the grain and then coarsely grinding the interior kernel. Barley groats (peeled barley, pearl barley) were used in soups. Oat groats (oat grits, oat meal[1]) and millet groats were boiled into a porridge. Buckwheat groats (grits or meal) were also boiled into a hot cereal, which was sometimes made into small cakes and fried.

Stamping Process. The first groat milling operation at the 1765 oil mill preceded the installation of the millstones. Millet was stamped into groats in 1765 and 1766 using the stamping mortars of the oilseed mill (figure 33),[2] a common Germanic practice of the period.[3] Barley groat production began early in 1766 using the same machinery.[4] This stamping process freed the hulls from the grain and cracked some of the kernels. The product was then cleaned by winnowing before use.

With the installation of the millstones on the top floor late in 1766, the stamping of barley groats ceased; but millet stamping apparently continued for some time.[5]

Grinding Process. As soon as the new millstones started operating,[6] groats were produced by the conventional five-step grinding process (figure 42). First the grain was cleaned by sieving and winnowing. Next the hulls were removed from the kernels by passing the grain through a pair of hulling millstones, and the loose hulls were blown away in a fanning (winnowing) machine. Oats required an additional drying step prior to hulling to make the hulls stiff enough for easy removal. The kernels were then coarsely ground into groats on a second pair of millstones, and the product was sieved into different sizes. The machinery used for grinding groats was quite similar to that found in mills grinding only wheat flour. However, it was more efficient for the Moravians to construct this

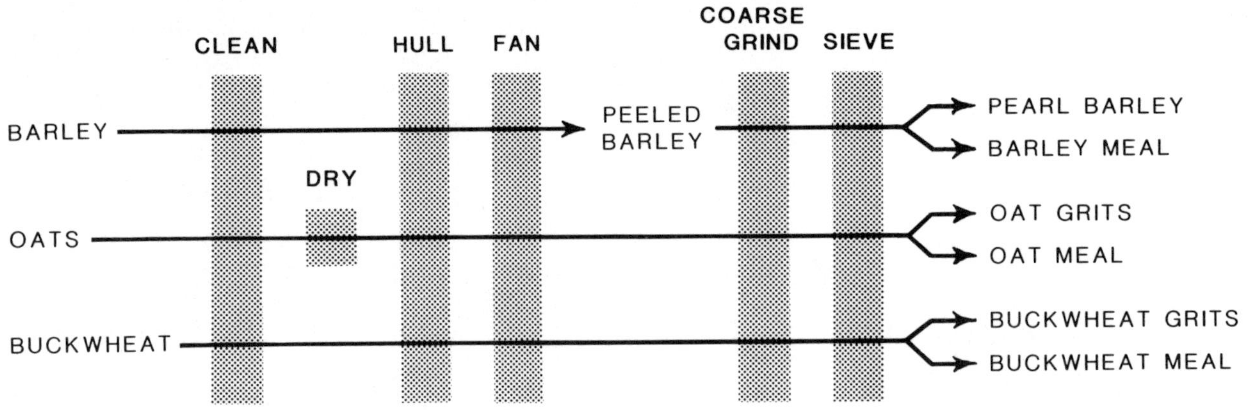

42. Groat milling at the oil mill was a five-step process, including two passes between millstones (first for hulling, then for grinding). Oats required an additional drying step to facilitate hull removal. Sieving separated the final product into various grades according to size.

specialized groat mill, allowing their nearby 1751 grist mill to cater to the increasing demand for quality wheat flour.[7]

The primary grains processed in the groat mill were barley, oats, and buckwheat.[8] The material received from the farm usually contained impurities that needed to be removed prior to milling. So the initial step in making groats was to remove unwanted dust, dirt, broken kernels, and coarse debris using either the fanning machine or the clapper mill.

The fanning or winnowing machine was located on the third floor and was run by waterpower.[9] Its exact design is unrecorded, but it was probably very similar to the typical machine depicted in figure 43. A continuous stream of grain from the hopper was passed through oscillating coarse sieves where larger impurities such as straw, leaves, and sticks were removed. Simultaneously a rapid stream of air was blown through the grain to remove the fines (dirt, dust, loose hulls). The grain fell from the coarse sieves onto an oscillating fine-mesh screen where smaller impurities such as broken kernels and foreign

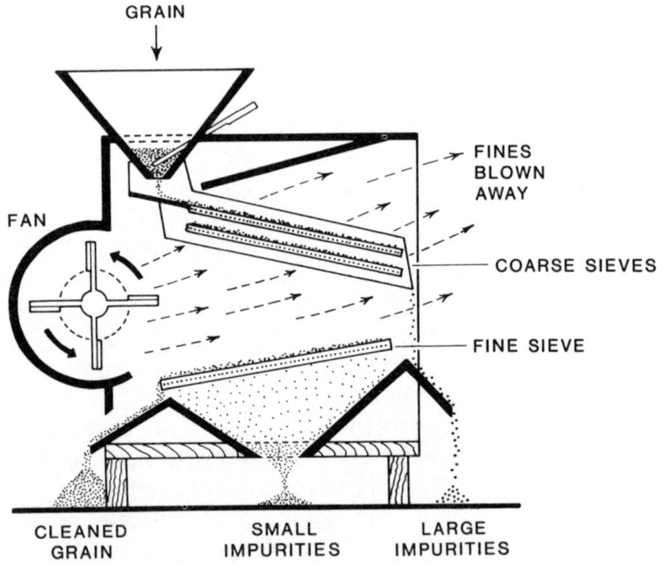

43. A fanning or winnowing machine was used to separate the grain from various impurities. The precise design used in the groat mill is unknown, but it was probably very similar to the typical fanning machine shown here in cross section.[10] A stream of air blown through the falling grain removed the unwanted fines. The grain also passed through and over oscillating screens to separate larger and smaller impurities.

seeds dropped through. The cleaned grain was collected as it fell off the end of this fine-mesh screen.

The clapper mill on the second floor was also used to some extent for cleaning grains.[11] Only the top view of this machine is shown in the 1766 Herrnhut drawings; a conjectural side view has been added based upon available evidence (figure 44). The clapper mill was a sieving machine in which an inclined tray having different size metal sieves was vibrated by six small trip-hammers.[12] Presumably the grain dropped onto the head of the tray, the fine fraction fell through the first sieve, the medium size fraction fell through the second sieve, while the coarse fraction dropped off the end of the tray. Each fraction was collected in the wooden boxes below. The camshaft activating the trip-hammers was powered by a rope drive from the upper camshaft of the oilseed mill (figure 28).

After cleaning, oats required an additional processing step. The oats had to be dried to make the hulls stiff and rigid so that they would be readily removed by the hulling stones.[13] This was accomplished by heating the grain for several hours in a special drying kiln located about forty feet (c. 12 m) south of the oil mill (figure 3).[14] The floor of the kiln was made of perforated iron plates covered with a hair cloth. The oats were spread to a depth of 4 to 6 inches (10-15 cm) on top of this cloth. Hot air rising from a stove with a long flue below the floor passed up through the iron plates and the grain and then out through a vent. The oats were turned frequently with a shovel to obtain even drying and to prevent scorching. When judged

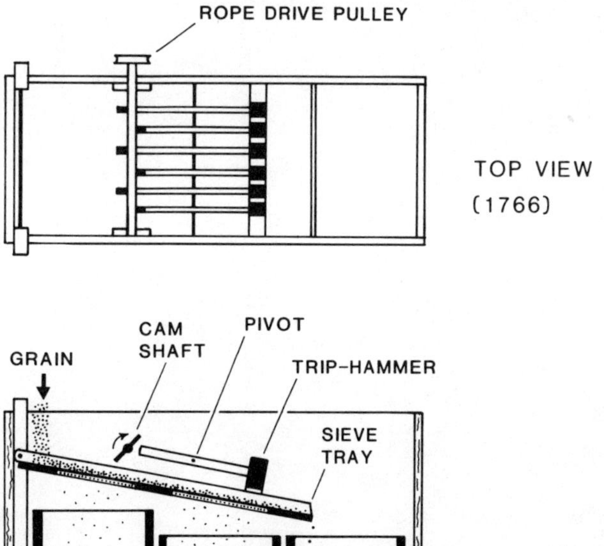

44. Grain was also cleaned and groats were sized using the clapper mill, a device having an inclined sieving tray vibrated by six small trip-hammers. The grain or groats flowed down the tray over progressively coarser metal sieves and separated according to particle size.

sufficiently dry, they were removed from the kiln and replaced with a new batch. Buckwheat was similarly dried on some occasions when the incoming grain was too damp.[15]

The grain was now ready for hulling. Barley, oats, and buckwheat are covered by non-nutritive hulls that are not lost during threshing and winnowing but must be removed before the kernel can be ground into groats. Hulling was accomplished by passing the grain between millstones set sufficiently far apart that the hull was scraped off but the kernel itself was not crushed. With barley the millstones were often set a bit closer so that both the hull and the outer skin of the barley kernel were ground off together. This setting produced "peeled barley" (*Graupen*), one of the two types of barley groats made at the mill.[16]

The 1766 Herrnhut drawings show housings for two run of millstones in the groat mill (figures 28-31). Few details of the actual machinery are given, however, probably because the drawings were made before the groat mill was completed.[17] A possible construction has been drawn in figure 46. The two millstones in each housing were 3.5 to 4.0 feet (1.07 to 1.22 m) in diameter. The gear train connecting the waterwheel to the rotating stones (figure 45) had an overall ratio of about 8:1, so the runner stones turned at approximately 60 revolutions per minute.[18] For hulling work, the millstones were usually

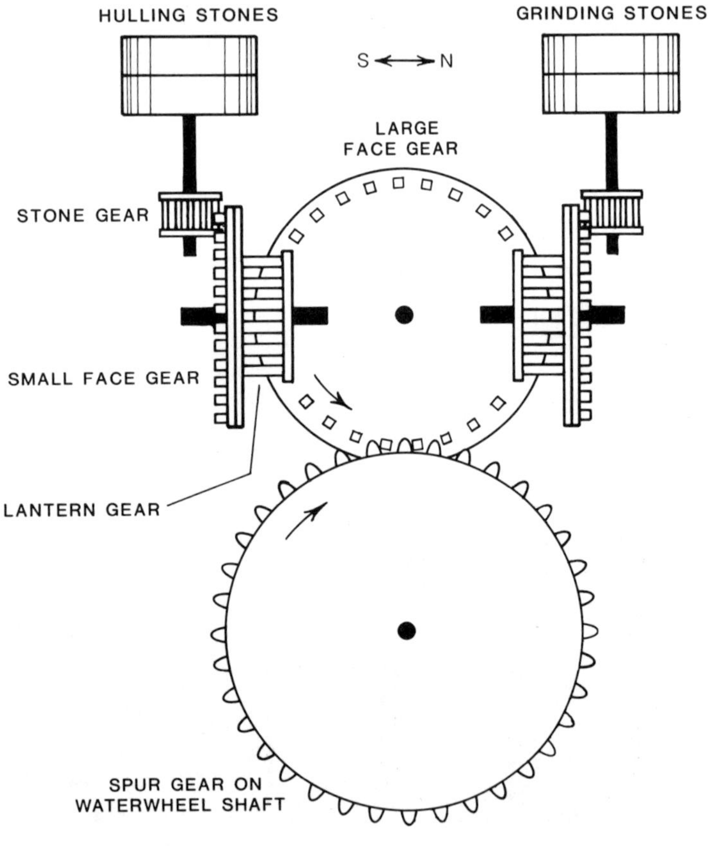

45. Schematic diagram of the unconventional three-step gear train used to drive the millstones in the groat mill. This system had high power losses due to friction but conveniently used the shaft of the large face gear as a camshaft for the northern stampers in the oilseed mill (figure 28).

46. A possible construction of the groat mill grinding machinery is shown in this cross-sectional drawing.[19] During operation grain fed from the hopper out a hole in the front of the vibrating shoe and then fell into the eye of the runner stone. Passing outward between the stones, the grain was hulled, cracked, or finely ground, depending upon how far the tentering adjustment lifted the runner stone above the bedstone.

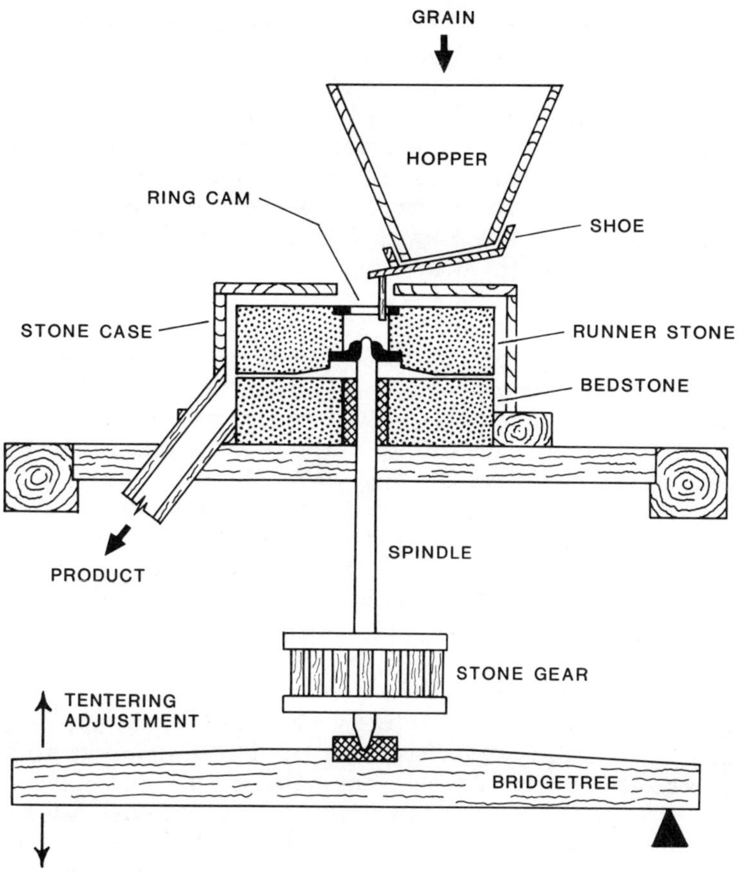

dressed with a sharp, rough surface having few or no furrows; and the upper stone often had a conical cavity cut around the eye of its grinding surface.[20] Since the stones for grinding groats had a different type of dress (see below), one run of stones (probably the southern) was dedicated to hulling work.[21]

Before beginning the hulling operation, the two stones were set approximately the length of a kernel of grain apart using the tentering adjustment (figure 46) which lifted or lowered the runner stone. During operation, grain from the hopper flowed out of a hole at the front of the shoe and fell into the eye of the runner stone. In most eighteenth century Germanic grist mills, grain flow was achieved by vibrating the shoe with a ring cam, which was set in the top of the runner stone and acted on a vertical rod attached to the bottom of the shoe.[19] As the grain moved outward between the stones, the sharp faces of the millstones rubbed off the hulls; but the kernel was not broken if the tentering was set correctly. The hulls and kernels then fell outside the stones and were swept into a chute leading to a collection bin on the floor below.

The product from the hulling operation contained kernels of grain, loose hulls, and a small amount of fines. This mixture was put through the fanning machine (figure 43) where the hulls and fines were blown away or sieved out. The hulled kernels were recovered and taken to the grinding stones, except for peeled barley which was now ready for sale (figure 42).

The grinding mill was nearly identical with that used for hulling (figure 46) except that the millstones were dressed differently. Stones for grinding groats generally had numerous furrows and the grinding surfaces between the furrows (the "lands") were given a finer dress than on hulling stones.[20] Before grinding the hulled grain, the clearance between the two stones was set with the tentering adjustment to produce the desired product; and this setting was readjusted as necessary during the run. For pearl barley (peeled barley ground further so that it was rounded like a pearl) a relatively wide gap between the stones was required.[22] For oat and buckwheat grits, a narrower setting was used. For oat meal and buckwheat meal, the stones were set very close together. The actual grinding operation proceeded very much like hulling. Grain fell from the hopper onto the vibrating shoe, then into the eye of the runner stone. Size reduction took place as the grain passed outward between the grinding stones, and the product fell down a chute to the floor below.

The product was then separated according to size using the clapper mill (figure 44). The location of the clapper mill almost directly beneath the grinding stones (figures 28 & 31) suggests that there may have been a connecting chute between the two operations, allowing continuous feed from one to the other.[23] In the clapper mill the grist slid down the vibrating tray sieve and dropped into the bins below. The separations obtained depended upon the sizes of the holes in the interchangeable metal sieves and varied from product to product.[11]

John Arbo sometimes recorded the yields obtained in groat mill operations during the late 1760's.[24] Ten bushels of barley yielded 5.0 bushels of peeled barley (*Graupen*) or 1.9-2.1 bushels of pearl barley. Ten bushels of oats produced 3.7 bushels of oat grits or 3.0-3.6 of oat meal, depending on how finely the oats were ground. Ten bushels of buckwheat yielded 4.2 bushels of buckwheat grits after hulling and grinding.[25]

Various other grains were sometimes processed in the groat mill. Spelt, a variety of wheat on which the husk remains after threshing, was husked from time to time using the hulling mill.[26] Maize (called *Welschkorn* by the Moravians) was occasionally ground into meal.[27] Large quantities of split peas were produced in 1768-1770,[28] and Arbo recorded a yield of 7.8-8.1 bushels of split peas from ten bushels of peas.[24] The groat mill also ground barley malt for the local brewery.[29]

Grain milling at the Bethlehem oil mill continued for over a century (1766-1875). Some equipment changes were made from time to time, but very little is known about these alterations. An oat meal sieving or bolting device was installed in the groat mill in 1794.[30] In 1814 the oat dryer was redesigned to reduce the fire hazard.[31] "A run of stones for buckwheat" was installed, possibly by

Charles D. Bishop who built up a thriving buckwheat flour business during his tenancy (1835-1847).[32] These may have been the French burr stones that were sold when the mill shut down in 1875.[33] No doubt Bishop also had a flour bolter for sifting his buckwheat flour. In general, however, the basic technology remained very similar to the original 1766 milling process.[34]

SNUFF MILL

The Moravians operated a profitable tobacco manufactory at Bethlehem producing cigars and snuff as well as smoking and chewing tobacco.[35] The bulk production of snuff required milling equipment to reduce the tobacco leaves to powder, and this processing was carried out at the Bethlehem oil mill from 1766 into the 1820's and possibly even later.[36] The oil mill did not make and sell finished snuff for its own account; rather it provided custom milling services for the Moravian tobacco manufactory and other local tobacco processors who made snuff.

The manufacture of bulk snuff was a complex procedure (figure 47). The tobacco leaves had to be cut up into small pieces, wetted with a salt solution, and fermented for a number of weeks in wooden bins to achieve the right flavor and texture. Then the tobacco was dried, milled, and sieved to remove any coarse particles. If desired, a flavoring solution was sprinkled on the milled tobacco; and after mixing, resieving, and aging, the snuff was packaged for the consumer. The entire process could be accomplished by hand labor, but mechanization of the milling step was highly desirable when large batches were prepared. Consequently many tobacco processors in and around Bethlehem contracted with the oil mill to dry, mill, and sieve the tobacco for their snuff.[38]

When the fermented tobacco arrived at the oil mill from the processor, it was usually too moist and soft for effective stamping or grinding. Hence the tobacco was first dried in the oil mill oat drying kiln (p. 59),[39] and the customer was charged for the firewood used.[40] The dried tobacco was then ready for milling.

Snuff milling at the Bethlehem oil mill falls into two definite eras. From 1766 to 1793 stamping machinery was used to produce a finely-cut tobacco.[41] Then in 1794 the Moravians purchased a new snuff mill capable of making a powdered product.[42] This latter mill remained in use for over thirty years.[36] Although tobacco milling was a major activity at the Bethlehem oil mill for six decades, no description of the precise machinery has been located. Nevertheless, comparison of the available facts with contemporary snuff milling practices in Europe and America provides a basis for estimating what the two snuff mills at the Bethlehem oil mill may have looked like.

Like other machinery in the 1765 oil mill, the tobacco stamping machinery used from 1766 to 1793 probably reflected contemporary Germanic technology. A typical

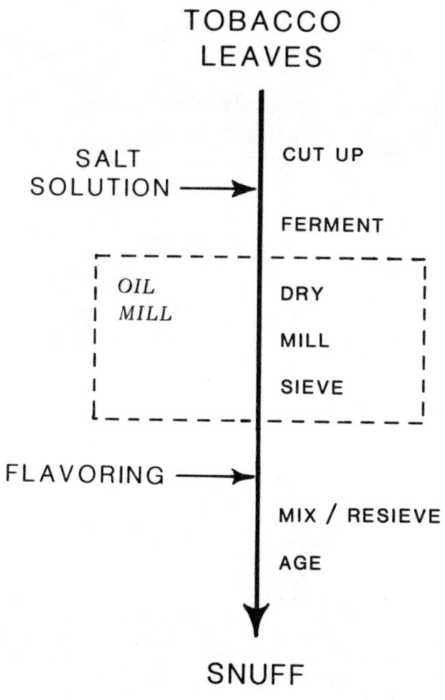

47. During the eighteenth and the early nineteenth centuries, the manufacture of bulk snuff usually involved seven major processing steps as outlined above.[37] Many local tobacco processors contracted with the Bethlehem oil mill to dry, mill, and sieve the tobacco for their snuff.

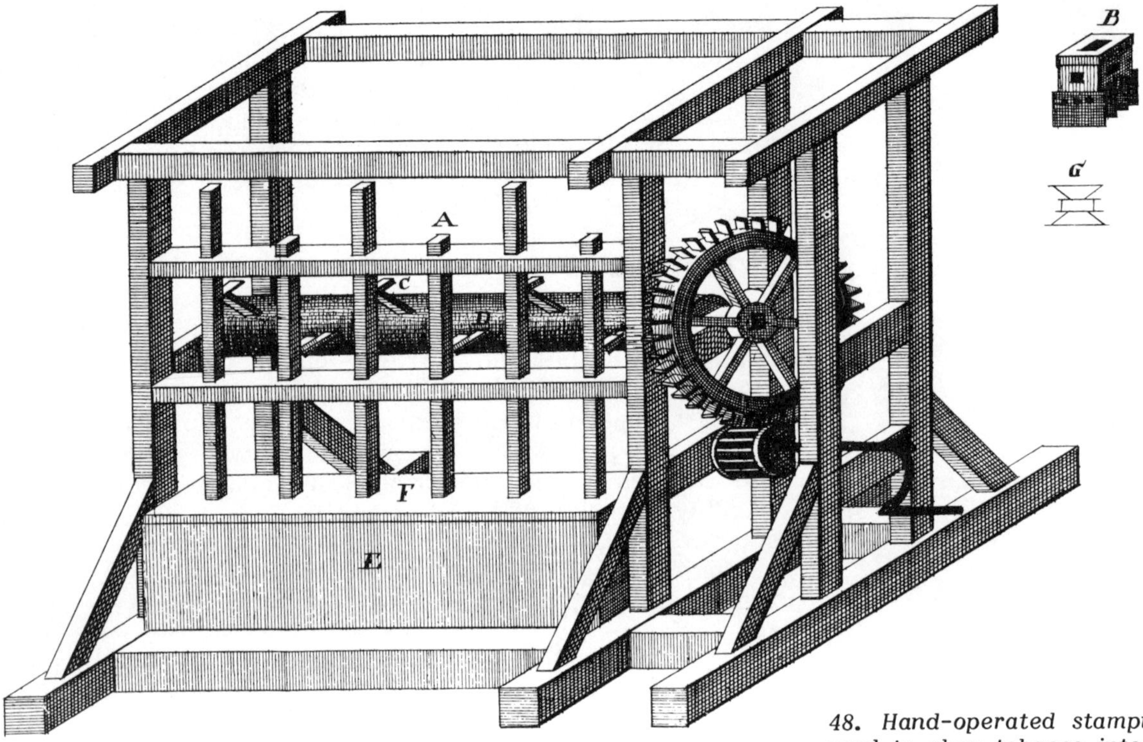

*48. Hand-operated stamping mill used to chop tobacco into finely-cut snuff in Germany during the late eighteenth century.[43] **A**, vertical stamps. **B**, stamping head with knives. **C**, lifting arm on stamper. **D**, axle with lifting cams. **E**, wooden stamping box. **F**, lid on stamping box. **G**, arrangement of ten cutting knives on stamping head.*

late eighteenth century German stamping mill used for producing snuff is depicted in figure 48. The tobacco leaves were placed in the closed box where they were chopped up by sharp knives located on the bottoms of the stamping heads. The six stamping poles, extending out through holes in the top of the box, were intermittently lifted and dropped by cams on axle **D** engaging the lifting arms **C** on the stampers. The mill depicted in figure 48 relied on the vibrations from the falling heads and also on occasional stirring to constantly shift the tobacco under the blades for even cutting; however, this was also accomplished in some snuff mills by mechanically moving the box during stamping.[44] Such a machine reduced tobacco leaves to a finely-cut product (not a powder) suitable for many types of snuff.

Tobacco stamping at the 1765 oil mill was probably done with the tanbark mill (figure 41), the only available stamping equipment constructed like the tobacco mill shown in figure 48. The tanbark stampers had the necessary cutting edges on the bottoms of the heads, and they were also enclosed in a wooden compartment to contain the tobacco dust stirred up during stamping.[45] It must have been necessary, however, to clean the tanbark mill thoroughly before stamping tobacco to prevent residual bark powder from adversely affecting the flavor of the snuff.

In 1794 a new tobacco mill was installed at the oil mill to grind powdered tobacco for snuff.[42] The reason for changing from the stamping mill that had produced finely-cut tobacco to the new mill that gave a product of flour-fineness is not clear. Perhaps powdered snuff had become more popular or maybe the new mill had a greater production capacity. The new tobacco mill was probably located in the western side of the oil mill building, since the oilseed and groat milling operations in the eastern side were quite active in 1794.

The design of the 1794 tobacco milling machinery is unknown. It was readily recognized as a snuff mill by outside visitors to Bethlehem,[46] so perhaps it was one of the common "snuff mulls" used in most Pennsylvania snuff mills during the 1790's.[47] These mulls generally consisted of a wooden mortar about 15 to 25 inches (c. 38 to 64 cm) in diameter with a weighted metal pestle (figure 49). Dry, cut tobacco was placed in the mortar and pulverized by turning the metal pestle with an offset drive arm. A heavy metal weight at the top of the pestle kept it in constant contact with the sides of the mortar as it rotated, but the pin connecting the pestle to the drive arm moved in a slot so that the pestle could roll over any lumps of tobacco. Later models of snuff mulls had a more cylindrical interior; the drive shaft extended down into the mortar and turned 3 or 4 grinding rollers bearing against the sides of the mortar.[49] In these later machines the ground tobacco was often removed from a gate at the bottom of the mull rather than scooping it out from above.

After stamping or grinding, the tobacco was sieved to remove any coarse particles, which were returned for

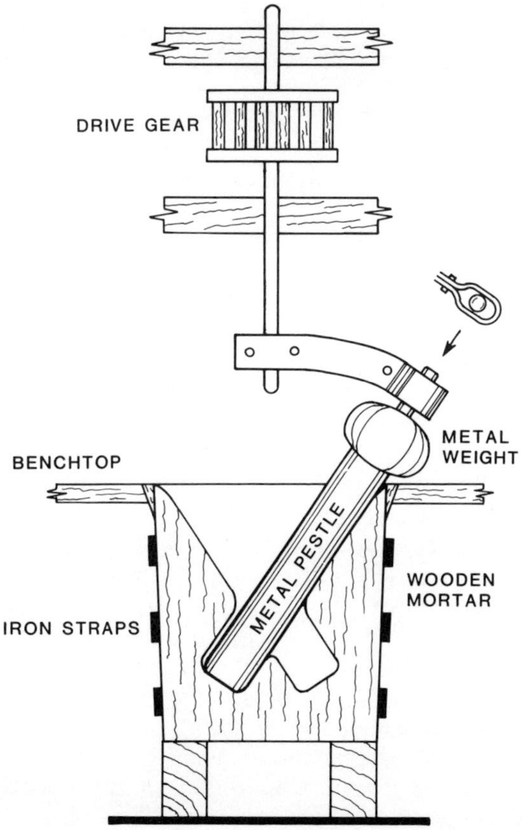

49. During the 1790's most Pennsylvania snuff mills ground their tobacco using a water-powered pot mull (mortar and pestle) of the general type shown here in cross section.[48] The new tobacco milling equipment installed at the Bethlehem oil mill in 1794 might have been of similar design.

further milling.[50] The finely-cut tobacco from the stamping mill could have been readily sieved on the clapper mill (figure 44) that was used for cleaning grains or for sizing groats. When the new tobacco mill was installed in 1794, a new bolter or sieve was also acquired to sift the powdered tobacco, but the exact construction of this device is unknown.[42] After sifting, the product was returned to the tobacco processor for conversion into the finished snuff (figure 47).

The size of the tobacco milling operation at the Bethlehem oil mill can only be roughly estimated from ledger entries. In the late eighteenth and early nineteenth centuries, the Moravian tobacco manufactory paid for about 500 to 1500 pounds (c. 225 to 680 kg) of tobacco to be ground each year at the oil mill,[40] but there were also other customers for this service.[38] From 1814 to 1821 tenant miller Gideon Helwig ground about 5000 pounds (c. 2300 kg) of tobacco annually.[36]

WATERWORKS

The oil mill served from 1832 to 1913 as the water pumping station for Bethlehem. During those 81 years various pumps and power sources were installed in the building (see table). Unlike Christensen's earlier pumps, most of this equipment was of contemporary commercial design, built to meet the pricing competition of the nineteenth century marketplace. Thus the successive installations reflected the changes in pumping technology during the period and the evolution of the power source from water (1832-1888) to steam (1868-1913) to electricity (planned for 1912).

Water Power. The initial pumping equipment was installed in 1832 in the eastern half of the oil mill where the oilseed mill had been located. The pump was powered by the existing waterwheel (figure 29). A new power transmission system connected the waterwheel shaft to the pump crankshaft through a clutch coupling.[51] This clutch provided a means of quickly disconnecting the pump from the waterwheel.

The 1832 pump was manufactured by Rush & Muhlenberg of Philadelphia. It is properly called a "simplex power pump" because it had only one cylinder and required a separate source of mechanical power. It was double-acting, meaning that it pumped water when the piston moved in either direction.[52] The pump had a piston diameter of 5 inches (12.7 cm), a stroke of 36 inches (91.4 cm), and was rated at 118 gallons (447 liters) per minute.[53] From these figures, it may be shown that the piston made 19.3 complete cycles every minute.[54] Since the waterwheel driving the pump had a maximum speed of about 7.5 revolutions per minute (p. 43), gearing or belting must have been used to increase the speed of rotation.

Pumping with waterpower was inexpensive but had one major disadvantage. Occasional high water in the Monocacy Creek caused water to back up into the wheelpit, slowing or stopping the waterwheel and shutting down the pump.[55] As a result, the waterwheel was replaced in 1846 by a Howd patent turbine to minimize such backwater shutdowns.[56] The performance of this turbine proved disappointing, however. The stroke of the pump had to be shortened and a flywheel installed before normal pumping operations were achieved. Ten years later in 1856, the turbine was replaced with another waterwheel, presumably similar to the original undershot design. Water-powered pumping continued until 1888 but it only served as a back-up for the steam pumps after 1868.

Steam Power. To keep up with the expanding water needs of the community, the waterworks turned to steam power in 1868. Two upright tubular Brinton & Henderson (Philadelphia) boilers with a combined rating of 40 horsepower[57] (30 kw) were installed, probably in the *Stube* and *Vor-Haus* area (figure 29).[58] They provided steam at a pressure of 70 pounds per square inch (4.9 kg per sq. cm) to drive the pump.

The Brinton & Henderson firm also supplied the new duplex, direct-acting, noncondensing steam pump installed in 1868.[59] A duplex pump (figure 50) is, in effect, two simplex pumps placed side by side, with the water entering through a common suction pipe and leaving through a common discharge line. To produce more uniform water delivery, the steam valves are interconnected so that one piston is in mid-stroke while the other is reversing direc-

CHRONOLOGY OF PRIMARY WATER PUMPING MACHINERY

PUMP			POWER SOURCE		
Date	Capacity	Type	Date	Power	Type
WATER POWER					
1832-1888	118 gpm* 447 lpm	Rush & Muhlenberg simplex power pump	1832-1846	c. 5 HP c. 3.7 kw	undershot waterwheel
			1846-1856		Howd turbine
			1856-1888		undershot waterwheel
STEAM POWER					
1868-1874	125 gpm 473 lpm	Brinton & Henderson duplex steam pump	1868-1874	40 HP 30 kw	Brinton & Henderson twin upright tubular boilers
1874-1913	350 gpm 1325 lpm	Cameron simplex steam pump	1874-1889		Noble & Rhoda twin horizontal boilers
1885-1913	350 gpm 1325 lpm	Worthington duplex steam pump			
1889-1913	1390 gpm 5260 lpm	Deane duplex compound steam pump	1889-1913	180 HP 134 kw	McKee & Milson twin horizontal tubular boilers
ELECTRICAL POWER					
1912**	1040 gpm 3936 lpm	De Laval two-stage centrifugal pump	1912**	85 HP 63 kw	Westinghouse electric motor

* gpm = gallons per minute, lpm = liters per minute, HP = horsepower, kw = kilowatts.
** Planned for the oil mill but installed in the new waterworks at the Illick's Mill site.

50. The Worthington steam pump installed in 1885 looked like this catalog illustration.[65] *It was a duplex design having two separate pumps mounted side by side (steam cylinders at left, water pumps at right) with coordinated operation of the two steam valves (top left). The bulbous air chamber on the water outlet (top right) cushioned the pulsations of the pump strokes to assure an even flow of water. The internal construction of this pump is shown in figure 51.*

tion. This was also a direct-acting pump, meaning that the steam piston and the water piston were at opposite ends of the same double-headed piston rod (figure 51). Such a design transfers the pumping force directly from the steam to the water, eliminating any revolving parts such as crankshafts or flywheels.[60] The new duplex pump delivered 125 gallons (473 liters) of water per minute in continuous operation. The steam cylinders were 12 inches (30.5 cm) in diameter; the water cylinders 6 inches (15.2 cm); and each piston had a 12 inch (30.5 cm) stroke.

The Brinton & Henderson boilers and pumps did not prove very durable, for the entire system had to be replaced after only six years. In 1874 two new horizontal cylindrical boilers were purchased from Noble & Rhoda of Allentown and installed in a new brick boiler house built at the southeast corner of the oil mill/waterworks build-

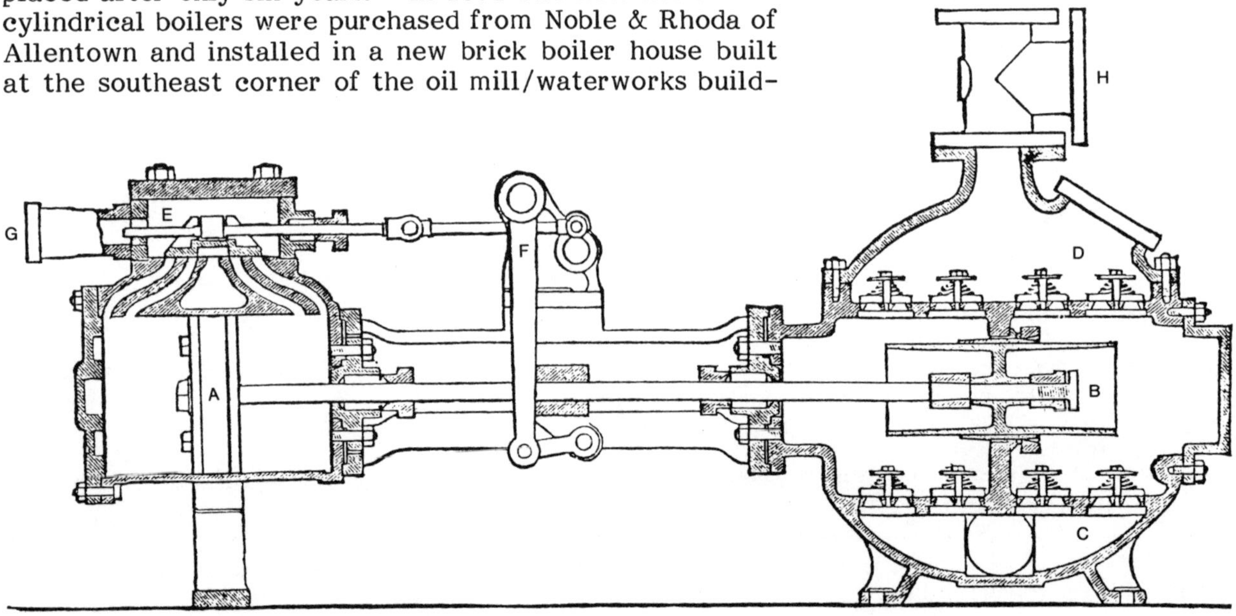

ing (figure 23).[61] These boilers were 3 feet (0.91 m) in diameter and 30 feet (9.14 m) long. Exhaust gases exited through a 45 foot (13.72 m) brick chimney (figure 24).

The new boilers supplied steam to a new, larger pump made by the Cameron Steam Pump Works of New York. This was a simplex, direct-acting, noncondensing steam pump with a pumping capacity of 350 gallons (1325 liters) per minute. The water cylinder on the pump was 12 inches (30.5 cm) in diameter; the steam cylinder had a 22 inch (55.9 cm) diameter; and both operated with a 36 inch (91.4 cm) stroke.[62] This simplex type of pump was the least expensive type to purchase and maintain but had a serious drawback. The sudden reversals of the pump piston at the end of each stroke caused severe shocks to the column of water in the discharge piping. If the pipes were long, this hammering could eventually rupture the pipes or valves. An air chamber was required for cushioning, with the trapped air acting as a spring to absorb much of the shock, then expanding as the pump piston reversed.[63] While all direct-acting pumps were usually operated with such an air chamber (figures 50 & 52), water pulsations were much more noticeable with a simplex than with a duplex pump.

Pumping capacity was increased in 1885 with the acquisition of a new duplex, direct-acting, noncondensing steam pump from the firm of H. R. Worthington (Brooklyn, New York). This new Worthington pump (figures 50 & 51) was powered by the existing boilers and delivered 350 gallons (1325 liters) of water per minute in normal service. The steam cylinders were 12 inches (30.5 cm) in diameter; the water cylinders 8.5 inches (21.6 cm); and each had a 10 inch (25.4 cm) stroke.[64]

During 1889 the waterworks again underwent a major renovation to increase its pumping capacity. New boilers and a new, larger pump were purchased; the interior of the building was rebuilt; the 1832 water-powered Rush & Muhlenberg pump was discarded; and the Cameron and Worthington steam pumps were retired to standby emergency service.

Two new horizontal tubular boilers made by McKee & Milson (South Bethlehem) were installed inside the main building in 1889.[66] They weighed 8 tons (7.3 metric tons) each and provided a total of 180 HP (134 kw), sufficient to run all three pumps (Deane, Worthington, and Cameron) simultaneously. Each boiler was 5.5 feet (1.7 m) in diameter, 15 feet (4.6 m) long, and contained 106 three inch (7.6 cm) tubes. A new 60 foot (18 m) iron chimney was erected to exhaust the boiler smoke.[67] There was also a separate boiler water feed pump[68] and a feed water heater and purifier.

A new 1390 gallon (5260 liter) per minute duplex, compound, noncondensing steam pump manufactured by the Deane Steam Pump Co. (Holyoke, Massachusetts) was also installed in 1889.[69] Not only was its capacity much larger than the combined capacity of the three pumps it replaced, but it made more efficient use of steam through its compound design (figure 52). This meant that there

*51. This sectional view from the Worthington catalog shows the double-headed piston rod used in the direct-acting type of steam pump.[65] When steam moved the piston head **A**, the motion was transmitted directly through the rod to the plunger **B** for pumping water. One-way valves allowed water to enter and exit the pumping chamber during the suction and pumping cycles. Water was pumped on both the forward and reverse strokes. **A**, piston in steam cylinder. **B**, double-acting plunger for pumping water. **C**, water suction chamber. **D**, water delivery chamber. **E**, sliding steam valve. **F**, lever arm operating steam valve. **G**, steam inlet. **H**, water exit. The external construction of this pump is shown in figure 50.*

were two steam cylinders. The first operated with high pressure steam, while the second utilized the low pressure steam exhausted from the first. All three pistons (two for steam, one for water) functioned in tandem on a common piston rod.[71] The Deane pump had 14 inch (35.6 cm) diameter water cylinders, 17 inch (43.2 cm) high pressure steam cylinders, 24 inch (61.0 cm) low pressure steam cylinders, and an 18 inch (45.7 cm) stroke.

In 1911 a local epidemic of dysentery and typhoid fever was traced to pollution of the spring from which the waterworks drew its water.[72] A device was installed for adding germicide (hypochlorite of lime) to the water as it was pumped, and a new pumping plant drawing from uncontaminated wells at the Illick's Mill site was hurried to completion.[73] As soon as it was ready, steam-powered water pumping at the oil mill/waterworks ceased in early January 1913.[74]

Electrical Power. Late in 1911 a two-stage, horizontal-shaft, 6 inch (15.2 cm) centrifugal pump with a pumping capacity of 1040 gallons (3936 liters) per minute was purchased to replace the aging Cameron pump. Manufactured by the De Laval Steam Turbine Co. (Trenton, New Jersey), it was driven by a Westinghouse 85 horsepower (63 kw), 220 volt a.c. electric induction motor.[75] Switching to an electric power system for pumping water promised to eliminate the need for operating a steam boiler at the waterworks and thus bring considerable cost savings. Although the new centrifugal pump and motor were intended for use at the oil mill/waterworks and were delivered there on December 7, 1911, they apparently were never installed at that site. Immediate concern with the contamination of Bethlehem's water supply dictated a rapid move to the new Illick's Mill waterworks location, so the De Laval pump was installed there instead sometime in September 1912.[76]

52. The Deane duplex steam pump installed in 1889 looked like this catalog illustration.[70] It had a still larger pumping capacity and utilized the more efficient compound design. This meant that each piston rod passed through two steam cylinders. The first (center) operated with high pressure steam, while the second (left) utilized the low pressure steam exhausted from the first. At the other end of each piston rod was the water pumping cylinder (right).

CONCLUSIONS

Several aspects of American technological development are illuminated by these descriptions of the 1766-1800 milling technology and the 1832-1913 water pumping facility at the Bethlehem oil mill.

Clearly Bethlehem was a "port of entry" for German technology coming to colonial America. The manufacturing techniques employed at the new 1765 Bethlehem oil mill were comparable to those employed at that time in central Europe. An immigrant millwright (Christensen) designed and built all the milling machinery, and he reproduced the technology that he and his fellow immigrants knew from their European homelands. The 1765 Bethlehem oil mill was not a unique example of imported eighteenth century Germanic technology in America, however; it is merely a case where extraordinary documentation has survived. Non-Moravian Pennsylvania-Germans of the period also operated similar mills, but the details of their processing machinery are largely unknown.[77]

The Germanic technology employed at the 1765 Bethlehem oil mill was different in many respects from the comparable British technology more commonly found in colonial America. Flaxseed was crushed in stamping mortars at Bethlehem rather than with the edgerunner millstones commonly employed in Britain.[78] Hemp was softened on a hemp roller mill at Bethlehem; the usual British practice employed a stamping mill.[79] Tanbark was prepared in a stamping mill instead of using the roller mill technique adopted in most British-American tanbark mills.[80] The initial Bethlehem snuff stamping mill reflected contemporary Germanic practice rather than the mortar and pestle type snuff mull generally used in Great Britain.[81] Of course many other aspects of the German-American and British-American milling technology were quite similar (the oilseed wedge press, for example), but it is clear that major differences existed.

Finally, the Bethlehem oil mill illustrates quite vividly how the source of milling and pumping technology was changing from the millwright to the manufacturer during the nineteenth century. The machinery installed there in 1765-1768 was all designed and built by the local millwright. Everything from the building to the waterwheels and from the gearing to the stamping mortars was constructed at the site under his direct supervision. Chistensen had built the 1762 Bethlehem waterworks in the same manner. On the other hand, when water pumping machinery was needed in the nineteenth century, it was purchased in the commercial marketplace. Technological expertise had become separated from the operational site, and the specialty manufacturer was shipping prefabricated machinery to his widely distributed customers.

CONSTRUCTION RECORDS FOR THE 1765 OIL MILL

The story of building the 1765 oil mill at Bethlehem comes mainly through four remarkable manuscripts: the "Oil Mill Diary," the "Oil Mill Account Book," the "Calculation of the Masonry at the Oil Mill," and the 1766 Herrnhut drawings (figures 28-31). These contemporary documents give a vivid, detailed account of the construction of the mill building and its machinery as well as the operation of that machinery. Without these sources of information, this book could never have been written.

Before writing our history of the Bethlehem oil mill, the "Oil Mill Diary," the "Oil Mill Account Book," and the "Calculation of the Masonry at the Oil Mill" were translated into English in their entirety. We present these translations here for the exceptional insights they give into mill building and technology in colonial Pennsylvania. This is the story of the 1765 Bethlehem oil mill in the words of the men who built it.

AUTHORSHIP

The authors of these three manuscripts did not identify themselves directly, but authorship can be ascribed with reasonable certainty through a process of deductive reasoning.

A detailed comparison of the handwritings in the three documents (figure 53) shows such a similarity of style that it seems likely that they were all written by the same person. That person was obviously interested in the day-to-day management of the oil mill construction project and its detailed financial expenditures. The author also must have been someone who enjoyed writing enough to keep a daily diary on the project over a period of years as well as to rewrite the financial accounts twice. Only two men could fit this description: Hans Christensen and John Arbo.

The evidence connecting Christensen with these oil mill documents is mainly of a negative nature:

53. Comparison of the handwritings in the "Oil Mill Diary," the "Oil Mill Account Book," and the "Calculation of the Masonry at the Oil Mill" manuscripts indicates that they were probably written by the same person.[1] The "Christensen" in the first two examples does not match Chistensen's signature (figure 54) but is identical with the "Christensen" written by Arbo (figure 55).

"OIL MILL DIARY"

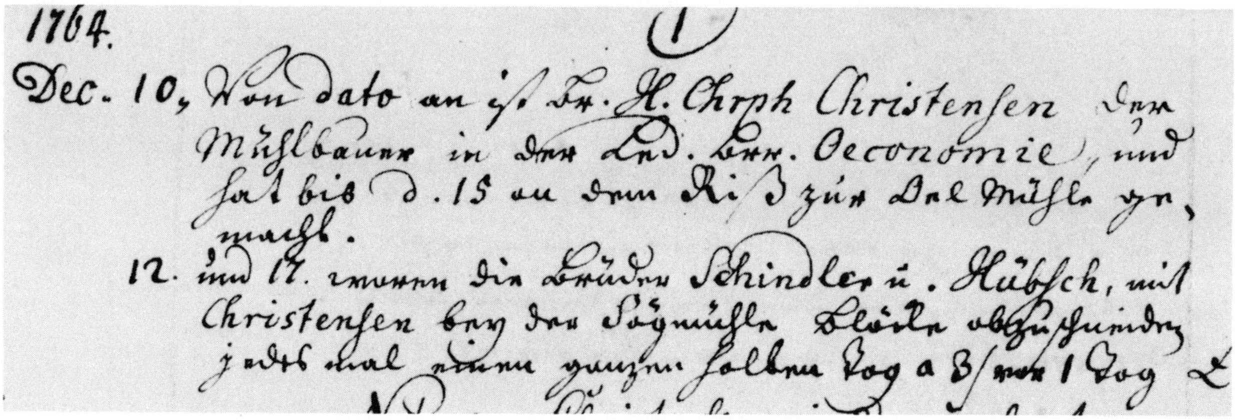

1764.

Dec. 10, Von dato an ist Br. H. Chrph Christensen der Mühlbauer in der [illegible] Oeconomie, und hat bis d. 15 an dem Riß zur Oel Mühle gemacht.

12. und 17. waren die Brüder Schindler u. Hübsch, mit Christensen bey der Sägmühle Klötze abzuschneiden jedes mal einen ganzen halben Tag a 3/ per 1 Tag £

"OIL MILL ACCOUNT BOOK"

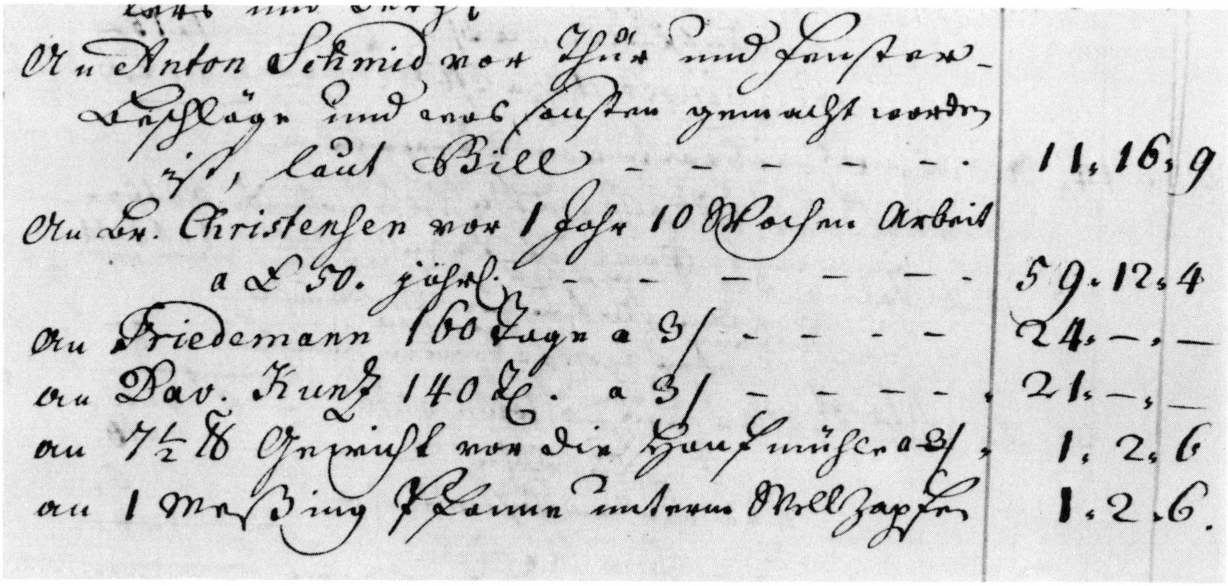

an Anton Schmied vor Thür und Fenster-beschläge und was sonsten gemacht worden ist, laut Bill - - - - - -	11, 16, 9
an Br. Christensen vor 1 Jahr 10 Wochen Arbeit a £ 50. jährl. - - - - - -	59.12.4
an Friedemann 160 Tage a 3/ - - - -	24.–.–
an Dav. Kunz 140 T. a 3/ - - - -	21.–.–
an 7½ lb Gewicht von die Hanfmühle a 3/	1, 2, 6
an 1 [illegible] Zinn [illegible]	1.2.6.

"CALCULATION OF THE MASONRY AT THE OIL MILL"

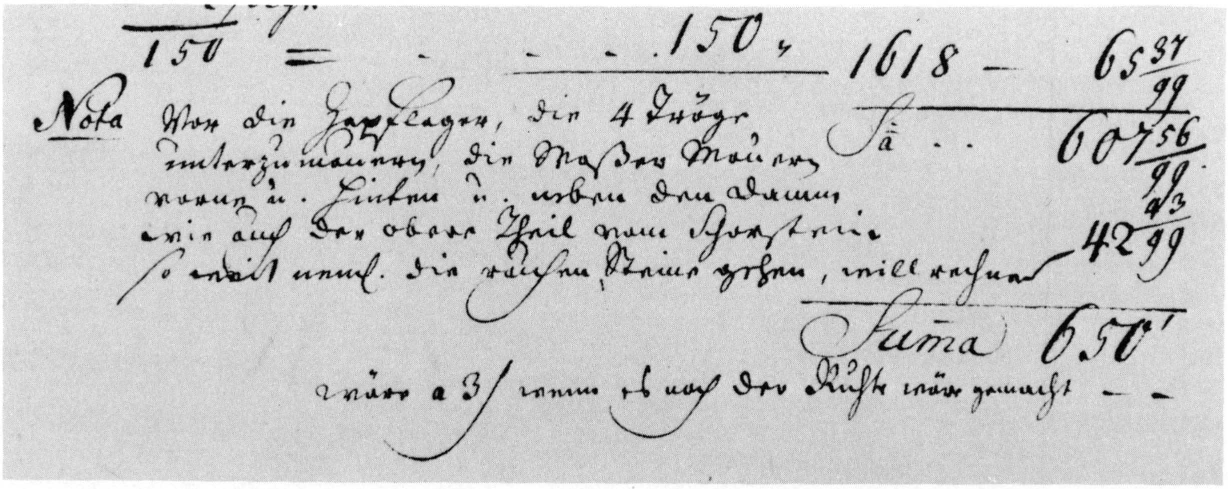

150 = . . . 150, 1618 – 65 37/99

Nota Vor die Zapflager, die 4 [illegible] unterzumauren, die [illegible]

fa . . 607 56/99 ; 4/3 ; 42 99

Suma 650

[illegible] a 3/ [illegible] [illegible] [illegible] [illegible] [illegible] gemacht - -

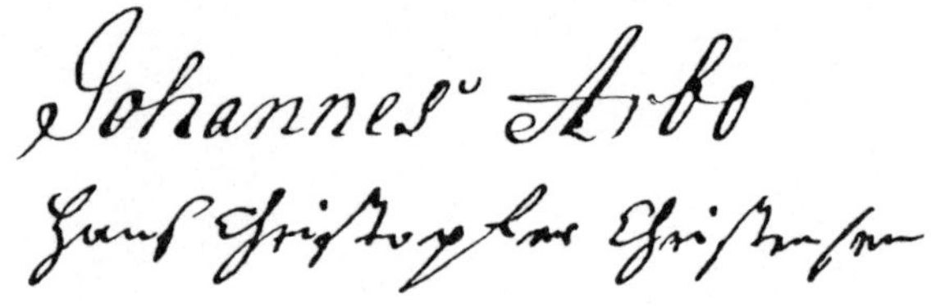

54. The signatures of Johannes (John) Arbo and Hans Christopher Christensen appear together on a 1762 document.[2] *Christensen's handwriting looks very different from that found on the oil mill construction manuscripts (figure 53), but Arbo's is quite similar.*

a) One entry in the "Oil Mill Diary" reads "Becker, Christensen, and I were in the forest searching for timber,"[3] clearly indicating that Christensen and the diarist were not the same person.
b) Only one unequivocal sample of Hans Christensen's handwriting is known, a signature on a 1762 document (figure 54). His style of handwriting is quite different from that found on the oil mill construction manuscripts.
c) An extensive search of the voluminous 1751-1776 Bethlehem records stored at the Moravian Archives has turned up no other documents definitely written by Christensen.[4] Apparently he was not a person who wrote extensively.

On the other hand, there is considerable positive evidence linking John Arbo to the authorship of these oil mill construction documents:

a) John Arbo's known handwriting (figure 55) bears a close resemblance to that found in the "Oil Mill Diary," the "Oil Mill Account Book," and the masonry calculations (figure 53).
b) Arbo's signature appears on page ten of the "Revised Oil Mill Account Book."[5]
c) The "Oil Mill Diary" contains many varied entries after 1768 when the oil mill was completed. These later pages are in the same hand as the 1764-1768 entries, but they end in 1772, the year of Arbo's death.
d) A 1765-1766 ledger account listing the building materials used for the oil mill was titled "*Oehl-Mühle Bau, John Arbo* [Oil Mill Construction, John Arbo]."[6] This clearly links Arbo with the financial management of the project.
e) The author of the "Oil Mill Diary" writes that he traveled to New York on July 8, 1766 and returned to Bethlehem on July 16. The "Single Brethren's Diary" reports that John Arbo left for New York on July 8, 1766 and came back on July 16.[7]
f) Arbo was a prolific writer. Many manuscripts have survived which can be attributed unequivocally to him.[8]

In light of the above, we conclude that John Arbo was the author of the "Oil Mill Diary," the "Oil Mill Account Book," and the "Calculation of the Masonry at the Oil Mill" manuscripts.

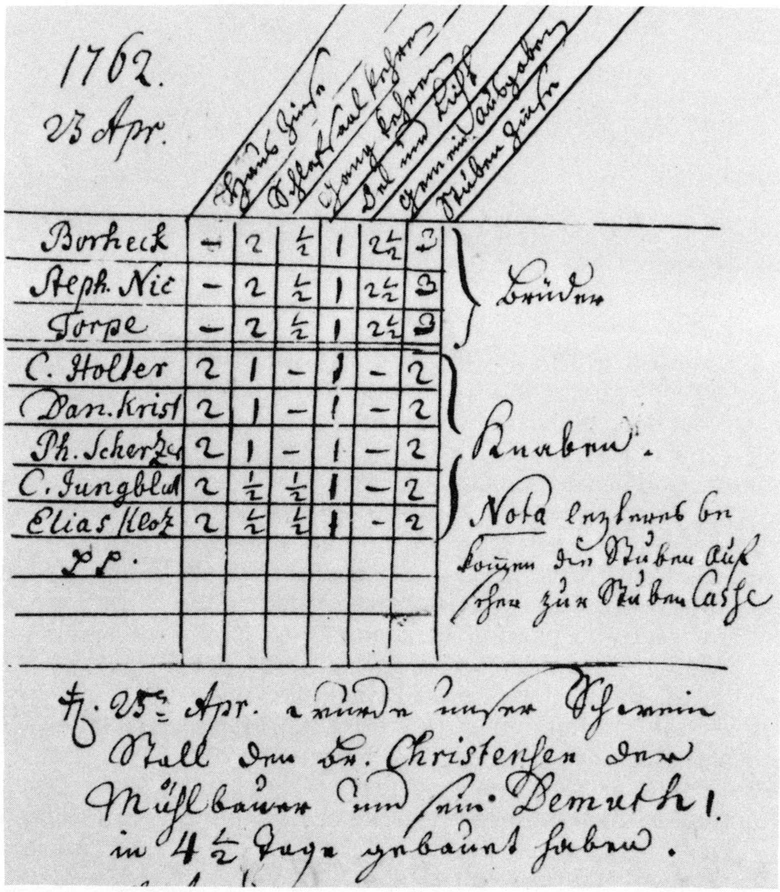

1762.
23 Apr.

Borheck	–	2	$\frac{1}{2}$	1	$2\frac{1}{2}$	3
Steph. Nic	–	2	$\frac{1}{2}$	1	$2\frac{1}{2}$	3
Torpe	–	2	$\frac{1}{2}$	1	$2\frac{1}{2}$	3
C. Holter	2	1	–	1	–	2
Dan. Krist	2	1	–	1	–	2
Ph. Scherzer	2	1	–	1	–	2
C. Jungblut	2	$\frac{1}{2}$	$\frac{1}{2}$	1	–	2
Elias Kloz	2	$\frac{1}{2}$	$\frac{1}{2}$	1	–	2

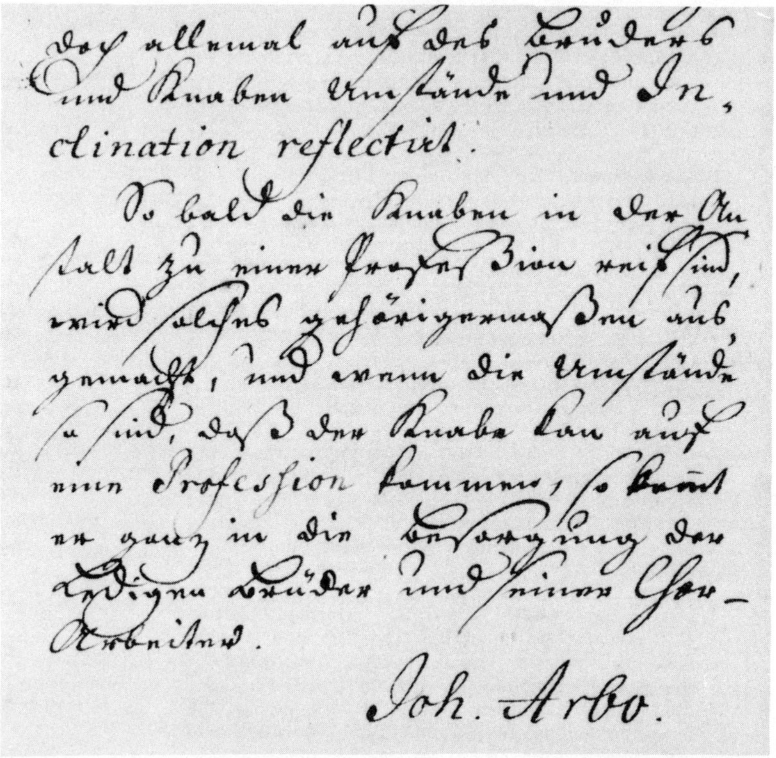

clination reflectirt.

Joh. Arbo.

55. *The handwriting in "John Arbo's Commonplace Book"* [9] *closely resembles the handwriting in the "Oil Mill Diary," the "Oil Mill Account Book," and the "Calculation of the Masonry at the Oil Mill" (figure 53). Note that Arbo mixes two different styles of penmanship: old German script for the main text and Latin script for names, foreign words, and emphasis. The same mixture occurs in the three oil mill documents.*

"OIL MILL DIARY"

The "Oil Mill Diary"[10] gives a fascinating day-by-day account of the building of the Bethlehem oil mill from Christensen's drawing of the initial plan in 1764 to the final completion of the hemp roller mill machinery in 1768. Diarist John Arbo adopted a very terse, factual style; some entries are all narrative, while others are mere financial accounts. Yet the flavor of the project and the people who participated comes through his words quite clearly. One reads of the perils of hauling building stones on sleds across the frozen Lehigh River, the elation when the rafters went up, the days taken off for fishing or for bleeding, and the joys of starting up the completed machinery.

The diary is not restricted to the construction of the new oil mill. Concurrent projects such as fixing pumps, boring and laying water pipes, or building an apple press sometimes took the workers away from the oil mill. Arbo records the time spent on these other projects so that he can accurately calculate the wages due for the oil mill work. Although the mill was completed in 1768, Arbo continued to record various other work schedules and business transactions in the book through 1772.

The English translation presented here runs from December 10, 1764 through February 5, 1768, the period during which the 1765 Bethlehem oil mill and its machinery were built. A complete translation of the remaining entries (1768-1772) is available at the Moravian Archives in Bethlehem, Pennsylvania.[11]

DECEMBER 1764

10 Starting today Brother H. Christoph Christensen is mill builder in the Single Brethren's Economy. Up to the 15th, he has worked on the plan for the oil mill.[12]

12 and 17 The Brethren Schindler and Hübsch were with Christensen at the sawmill cutting logs. A full half day each time @ 3/- per day. £-.6.-

Note: Christensen's time is calculated at 4/- per day plus 1/6 for board, which comes to 5/6 daily during the summer. (Has been changed. He gets £50.-.- yearly.)[13]

17 M Becker, Christensen, and I were in the forest searching for timber. £-.2.-

18 Tu Christensen's inventory was taken.

19 W Christensen was in the forest with Becker searching for timber. Schindler and Möhring helped load lumber at the sawmill. £-.2.-

20 Th Christensen and Straus were in the forest felling timber. Quite a snowstorm came during the night.

21 F Same men in the forest cutting timber. The first 4 logs were hauled through the snow to the meadow.

22 Sa Same men cutting timber in the afternoon. Hauled all day.

31 Up to today Christensen has organized his workshop and put his tools in order. During this time the masons have quarried stones behind the old lime kiln. Brother Schenke has quarried stones for me at the Monocacy [Creek] below Hirte's house, at 3 places along the Lehigh [River], and at the fence corner by the old lime kiln. For this I paid him £1.12.-.

JANUARY 1765

2 Since the snow was so deep and since the Lehigh above the large island was frozen so hard that we could go across it yesterday, the following preparations were made today to haul stones from Astrom's hill:[14]

a) The masons Schober, Schenke, Gerstberger, and Franz Seifart went across to straighten up the stone quarry at Astrom's hill and to roll down stones. @ 3/6

b) Claus Cölln, Just, and Hornig started to cut out a path on the bank of the Lehigh on this side so that one can load a horse sled there and drive away. @ 3/-

c) Brother Kliest made ice spurs, like the fishermen have in Germany, for those who will haul stones.

d) Brother Huber put prongs on 4 pikes for the stone haulers.

e) Christensen and Schindler made two handsleds for the stone haulers,

f) and Anton Schmid sheathed wooden shovels for removing snow.

At 3 o'clock in the afternoon the sleds were ready. Claus Cölln from Lititz, who was snowed in here, Möhring, Schindler, Hornig, and Christensen began to haul stones across the Lehigh to this side where they were picked up with large sleds and hauled across the meadow.

3 Th Schober, Schenke, Gerstberger, and Franz Seifart at the stone quarry. Hübsch, Claus Cölln, and Schindler cutting path. Each 1 day, except Schind-

Date	Day	Entry	Masons' Quarry	Schenke	Astrom's Hill
			Sledloads of Stones[15]		
		ler ½ day. Christensen, Just, and Hornig hauling stones. Hauled with 1 sled during the afternoon.			3
4	F	The same Brethren again today; 2 sleds for hauling stones. Today and yesterday, 1 quart of rum and 12 pretzels each day. Danielsen and Jorde helped during the afternoon, as did John Bruce.			14
5	Sa	All who were there yesterday were outside; but they had to stop in the morning since it snowed. Every day 1 quart of rum, and 1 quart on the 3rd. It snowed all day today.			6
6	Su	Still snowing the entire day.			
7	M	It stopped snowing. This latest snowfall was 14" deep. Together with the earlier snowfall, which had already settled some, the snow was 30" or 2½' deep. So we had to shovel out a path across the Lehigh to be able to haul stones again. Two sleds hauled in the afternoon. The following Brethren were hauling and quarrying stones: Schenke, Schober, Gerstberger, Franz Seifart, Just, John Bruce, Rauschenberger, Straus, Hübsch, Claus Cölln, and Christensen.			6
8	Tu	The same men were there again, as were Wm. Volk, Hornig, and Danielsen for ½ day. Two sleds hauled.			12
9	W	All of yesterday's Brethren were there. Three sleds hauled.			18
10	Th	Today the same Brethren were at the stone quarry. Since the weather was mild, there was water on the ice beneath the snow, which made the handsleds difficult to move. Three sleds hauled the entire day.			18
11	F	During the night there was a mild frost, and it started to snow again. Therefore, no stones were hauled across the Lehigh today, but all the Brethren rolled down stones. Although 3 sleds were used, each made only 2 trips because it snowed more and more. During the afternoon everyone stayed home.			6
12	Sa	There was a hard freeze again during the night, and the wind was from the northwest. Christensen, Straus, Just, and Rauschenberger widened the path across the Lehigh. The sleds were sheathed with iron runners, and new boards were mounted on them; but nothing was hauled today.			
14	M	Stones were quarried and hauled again.		3	15
15	Tu	Same.			18
16	W	Since the sun shone so strongly and the weather was so mild yesterday, the ice on the Lehigh was weakened, especially since so much snow lay on top of it. Therefore, we had to stop bringing stones across this morning. So this afternoon the masons came on this side of the Lehigh to quarry stones. Today we hauled	6		12
17	Th	Early in the morning we went upstream by canoe and found that a quarter of the ice on the Lehigh had already broken up, particularly in the center of our path as far as Jennis Pond. Therefore, we broke up a lot of the ice, continuing this work throughout the afternoon. We took up the flatboat to use for hauling stones across in the morning.	15		
18	F	It had already started snowing early so we could not transport [any stones]. The snowstorm lasted the entire day. Toward evening it started to storm from the northwest and continued the entire night.			
19	Sa	It cleared up early but the wind continued. Because the water might rise or the ice break up, we did not want to haul [any stones] for the time being. So the flatboat was brought back down again.			
21	M	In the afternoon Brother Just went over to Astrom's hill to roll down all the stones that were already quarried.			
22	Tu	The path from here to the ferry was shovelled open. Today we hauled stones again from the stone quarry at the old lime kiln using 3 sleds.	18		
23	W	Were hauling stones to the oil mill again in the morning.	9		
24		From the masons' stone quarry.	5		
		From the Lehigh at the landing.[16]			4
		Sum	53	3	132
		Sled[loads to date]			188
28	M	Since yesterday and today were as cold as New Year's, Brother Just went up the Lehigh early to see if it was frozen over yet. He brought back the news that one could haul [stones across it]. So 6 Brethren left before 9 o'clock and made 14 trips with the handsleds during the morning. The teamsters were also called in and hauled with 2 [large] sleds during the afternoon.			4
29	Tu	They hauled with 3 sleds.			18
30	W	They hauled with 3 sleds again.			18
		Sum			228
		During the afternoon it started to snow again and continued until			
31	Th	the afternoon. So no stones were hauled either over there or over here.			

FEBRUARY 1765

8 The weather turned mild in the morning so it was decided to bring up all the rest of the stones that had been left on this side of the Lehigh. There were 12 sledloads

altogether plus one sledload we still had at the old lime kiln. Thus a total of 13

[Total sledloads of stones] 241

In the afternoon it rained and thawed very fast.

19 We brought the flatboat from the Lehigh into the Monocacy to carry stones.

20 Since it was frozen hard, we brought our two shaft logs out of the forest.

23 We cleaned the Monocacy because we wanted to bring up stones with the flatboat.

28 Th Started to haul [stones] once again after many of the Brethren had been sick. Möhring, Bruce, and Hübsch were there, and during the afternoon Just.

MARCH 1765

1 F Peter Krist, Bruce, Just, Hübsch, and Jorde were there.

2 Sa We brought up the largest stones. Christensen, Rauschenberger, Schenke, and Peter Krist were present.

4 M Christensen and Rauschenberger started to square logs behind the stable. Starting today Brother Rauschenberger is with Christensen.

7 Th Christensen, Rauschenberger, Peter Krist, and Wm. Volk went up to Thomas Bits' island and fetched the two chestnut logs left there a year ago. They split each log in half, fastened them one behind the other with vines, and brought them downstream with the canoe.

8 F In the morning the same four Brethren brought the logs onto land. During the afternoon Christensen, Rauschenberger, and Peter Krist sawed off [pieces] and split them for [making] shingles.

9 Sa Christensen, Rauschenberger, Peter Krist, and Wm. Volk finished splitting the chestnut logs. Joh. Seifart hauled them up.

11 M Started to split shingles. Christensen, Rauschenberger, and Peter Krist, 1 whole day. Wm. Volk, afternoon.

12 Tu Above four Brethren split shingles again.

13 W and 14 Th The four Brethren continued making shingles, finishing on the 14th.

15 F and 16 Sa The pipes for the Single Sisters and for the pump in Emmaus were brought up from the Lehigh. The above 4 Brethren started to bore.

18 M Christensen, Peter Krist, and Rauschenberger finished with the pump boring.

19 Tu Same three Brethren started to bore the pipes for the Single Sisters.

20 W They completed the work in the morning and started to lay the pipes in the afternoon.

21 Th The above three Brethren were still laying pipes at the Single Sisters until 3 o'clock. Then they went to the Lehigh to cover the pumps.

22 F There were high winds and snow the entire night. The snow continued today until late at night.

23 Sa They squared the main shafts.

26 Tu Christensen, Rauschenberger, and Peter Krist started to square logs at the meadow.

27 W They had to stop while Brother Christensen hurried to Lititz to stake out the site for the fulling mill and the dam.

31 Su Brother Christensen returned, and

APRIL 1765

1 M they squared logs at the meadow again.

2 Tu Same.

3 W Same.

4 Maundy Thursday
5 Good Friday
6 Holy Saturday
— There was no work.

9 Tu Christensen, Peter Krist, and Rauschenberger squared logs at the meadow again. They continued every day until

13 Sa they finished.

15 M Christensen, Peter Krist, and Rauschenberger went to the Lehigh to work on the pumps, remaining until noon. In the afternoon they worked on the stamping troughs behind the stable.

16 Tu They worked behind the stable again, Peter Krist and Rauschenberger, that is. Christensen bored pumps at the Lehigh.

17, 18 They continued up here. Brother Christensen came up here the second afternoon and completed the valves, etc. for the pumps. On

19 and 20 he helped the other Brethren working on the stamping troughs.

22 M They worked on the stamping troughs again.

23 Tu Christensen completed Georg Rothrock's pump, which Rothrock picked up this evening.

24 W In the morning Christensen and Rauschenberger went to the Settlement[17] to install the pump. They returned at noon and worked on the stamping troughs.

25 Th Peter Krist and Rauschenberger worked on the stamping troughs. Christensen went with Becker into the forest to search for some more timber.

26 F Today they bored out the brass barrel cast by Tommerop for the Emmaus pump.

27 Sa They installed the barrel in the morning. In the afternoon Peter Krist and Rauschenberger repaired the fish dam because they want to fish this coming Monday.

29 M They worked on the Emmaus pump again.

30 Tu All 3 went fishing.

MAY 1765

1 W It rained, so they could not work in the morning. They started work again in the afternoon.

2 Th They went fishing again.

3 F The Emmaus pump was picked up, and Christensen went to install it.

4 Sa Christensen was in Emmaus. The other two took a holiday today because they were tired from fishing.

6 M Christensen went to Emmaus to do something with the pump. The other two made wooden pegs.

7 Tu Christensen returned at noon. They continued [making pegs], but the last 3 days they went into the forest to search for curved timber for the waterwheels.

13 M and 14 Tu They cleaned up back in the yard and below along the Monocacy. The lumber was moved to one side so that one could drive a wagon in both places. The lime shed[18] was started.

15 W They went fishing.

16 Th Ascension Day was celebrated.

17 F and 18 Sa They went fishing again.

20 M Christensen left early for Lititz. Peter Krist repaired the flatboat. Rauschenberger had a bad hand and could not work.

21 Tu Peter Krist worked alone on the flatboat, and Rauschenberger cleaned chimneys.

22 W Peter [Krist] and Rauschenberger went fishing.

23 Th They worked on the flatboat and the lime shed.

24 F They hauled lumber from the sawmill with the flatboat.

25 Sa Fishing.

27 M It was Whitmonday, a holiday.

28 Tu and 29 W The lime shed was sufficiently complete that lime could be hauled in. 2 days, 12/-

30 Th They started to build the pigsty.

31 They continued.

JUNE 1765

1 Sa The same.
3 M, 4 Tu The pigsty was completed.
5 W They worked on the lime shed again.
6 Th They hauled lumber from the sawmill with the flatboat. — 2 days, 12/-
7 F Our meadow was mowed.
8 Sa They worked on the pig trough.
10 M They completed the pig trough in the morning.
11 Tu The pigsty was totally completed by breakfast. Then Peter [Krist] and Rauschenberger went to clear the island. In the afternoon Rauschenberger helped Boeckel loading hay.
12 W Rauschenberger was with Boeckel the entire day, so Boeckel gave us Brother Weigand in his place again.
13 Th, 14 F They worked on the island.
15 Sa Rauschenberger was alone because Peter [Krist] wanted to go for bloodletting. From the 17th to the 22nd they both worked on the island and in the garden. Rauschenberger cleaned chimneys during the last days.
24 M Since Christensen had returned from Lititz last Saturday the 22nd, today we started to dig [a new channel] at the Monocacy. This will divert water from the building site where the oil mill is to stand.[19] The following Brethren were there: James Hall, Just, Peter Krist, and Rauschenberger. Christensen completed the pumps to pump out the water [from the building site] into the Monocacy.
25 Tu The above four Brethren were again digging to draw off the Monocacy. Christensen was at the sawmill the entire day to cut curved wood.[20]
26 W James Hall was digging only until breakfast, but we got Weigand in his place. Christensen was at the sawmill.
27 Th Just, Peter Krist, Rauschenberger, and Weigand were digging; and Christensen was at the sawmill again.
28 F The above four Brethren were digging. Christensen was still at the sawmill in the morning and helped with the digging in the afternoon.
29 Sa All five were digging and making the dam.[21]

JULY 1765

1 M Christensen, Rauschenberger, Just, and Weigand worked on the dam again.
2 Tu Everyone from yesterday was present except Just.
3 W Just, Peter Krist, Rauschenberger, Weigand, and Melchior Krist worked making the dam.
4 Th Christensen, Peter Krist, and Rauchenberger were at Georg Rothrock's to install his pump. Weigand could not work because of a bad finger, and Just was being purged. So Melchior Krist was all alone.
5 F Rauschenberger, Melchior [Krist], and Peter Krist cleared away the old foundation.[22] Christensen repaired the waterworks.
6 Sa Started moving away the [old] waterwheel.[23]
8 M The waterwheel was brought to the site where it will remain until it can be set again on its bearings. Rauschenberger, Peter Krist, Just, Melchior Krist, and Andrew the Black worked on cleaning and clearing the ground today.
9 Tu Christensen was ill. Those Brethren who worked yesterday continued today.
10 W Rauschenberger, Peter Krist, Melchior Krist, and Andrew the Black worked. Just went fishing.
11 Th Just and the others were all there.
12 F Rauschenberger, Peter Krist, Melchior Krist, Andrew the Black, Sturgius, Just, Schenke, Gerstberger, and Jacob Ernst were all there. At 4:30 this afternoon the cornerstone was laid, and they started building the walls. Bailing and pumping continued all night.[24] The [young] boys pumped until 10 o'clock; from midnight on it was Sturgius, Feldhausen, Pell, and Beverhaut.
13 Sa Rauschenberger, Peter Krist, Mechior Krist, Just, Andrew the Black, Schenke, Gerstberger, Ernst, and Sturgius were there working again. In the evening the boys pumped until 10 o'clock, and Starck, Jorde, Jacob the Mulatto, and Hartmann from 2 o'clock until 6.
14 Su The Brethren and the boys continued pumping and bailing the entire day. Feldhausen, Dreysprüng, and Tommerop pumped from 2 o'clock at night until 5 in the morning.
15 M Rauschenberger, Peter Krist, Melchior Krist, Just, Andrew the Black, Schenke, Gerstberger, Jacob Ernst, and Sturgius were at work today. Up to now Christensen with Rauschenberger and Peter Krist helped dig the foundation; but tomorrow they will go to their lumber work. Feldhausen, David Beck, and Tommerop pumped from 2 to 5.
16 Tu Sturgius, Melchior Krist, and Andrew the Black worked digging the foundation. Gerstberger, Schenke, and Jacob Ernst did masonry work. Christensen with Rauschenberger and Peter Krist did lumber work, and Schindler also helped. Feldhausen, Pell, and David Beck pumped from 2 to 5 o'clock.
17 W Sturgius, Melchior Krist, Andrew the Black, and Just were digging. Gerstberger, Schenke, and Jacob Ernst did masonry work. Christensen, Rauschenberger, Peter Krist, and Schindler did carpentry work. Feldhausen, Tommerop, and David Beck pumped from 2 to 5 o'clock.

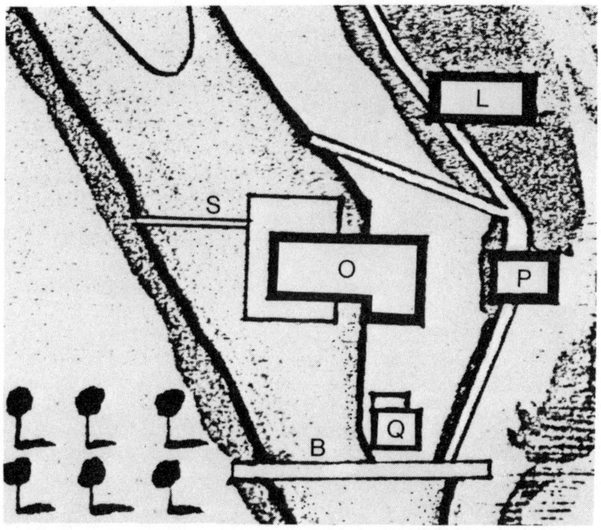

56. This detail from a 1772 map of Bethlehem (figure 3) records the site plan of the 1765 oil mill shortly after it was built. The June, July, and August 1765 entries in the "Oil Mill Diary" relate how a cofferdam was built and the course of Monocacy Creek was relocated c. 33 feet (10 m) to the west so that the western half of the oil mill foundation could be laid in the former creekbed. ***B****, bridge.* ***L****, tanbark shed.* ***O****, oil mill.* ***P****, waterworks.* ***Q****, oat drying kiln.* ***S****, dam.*

18 Th Everyone from yesterday was at work again. Feldhausen, Pell, and Dreysprüng pumped from 2 to 3 [pm]; and in the evening the boys pumped until 9:30. Feldhausen, Pell, and Dreysprüng pumped from 2 to 5 o'clock.

19 F The same Brethren as yesterday were all at work again. The sill for the sluice gate was laid under the raceway [forebay] today. Feldhausen, Tommerop, and Beverhaut pumped from 2 to 5.

20 Sa Only Just was absent today. Feldhausen, Pell, and Hartmann pumped from 2 at night until 6 in the morning.

21 Su The boys pumped until 9 o'clock. After that Feldhausen, Dreysprüng, and Matthiesen from 2 to 5 in the morning.

22 M Rauschenberger, Peter Krist, Just, Andrew the Black, Schenke, Gerstberger, Melchior Krist, and Jacob Ernst were at work today. Brother Schenke also got the helper that he has had before, namely Joh. Knecht. Knecht will take Ernst's place but remain a helper when Ernst goes to Wachovia.[25]

23 Tu All the Brethren from yesterday were at work again. This morning it started to rain a little.

24 W Rauschenberger, Peter Krist, Melchior Krist, Just, Andrew the Black, Schenke, Gerstberger, Jacob Ernst, and Joh. Knecht were at work. Because it started to rain in the afternoon and there was no hurry to finish digging, Melchior Krist went to his workshop, Andrew the Black also went home, and Just called it a day too. The Brethren Just and Carl Holter agreed to take care of the pumping for the entire night for 4/-. They started this night.

25 Th Only the carpenters worked. Christensen fixed the pump in the laundry house. Since it was still raining a little, the masons went to the bakery and started to whitewash. But Schenke and his helper remained at the pump. Just and Holter pumped during the night.

26 F The masons and carpenters were at work again. Just was there for ¾ of the day, i.e. from nine o'clock on.

27 Sa Rauschenberger, Peter Krist, Schenke, Gerstberger, Jacob Ernst, and Joh. Knecht were at work. Just, ¼ day. After breakfast the two walls for the gudgeon bearings in the raceway [wheelpit] were built. Today was Communion. Those boys who did not attend pumped in the evening.

28 Su Pell and Jacob the Mulatto pumped from 2 to 6 o'clock. During the day and evening the boys and a few Brethren who wanted to did the pumping. Just and Holter pumped during the night.

29 M Christensen with Rauschenberger and Peter Krist were fitting together planks for the raceway. Thomas also helped them. Schenke, Gerstberger, and Jacob Ernst on the masonry work. Joh. Knecht and whichever Brethren came down pumped. Just and Carl Holter started hauling away the dam that had been put up between the old and new courses of the Monocacy across to the meadow. They worked only ½ day and took care of the pumping during the night.

30 Tu Everyone who worked yesterday was also here today. Just and Holter, ½ day and pumped at night.

31 W All the Brethren from yesterday were working today. Schindler and Thomas helped the entire day. Just did not work today, but he and Holter took care of the pumping during the night.

AUGUST 1765

1 Th All the Brethren from yesterday were working except Thomas. Jacob Ernst, who is going with Jacob Steiner to Wachovia, finished working today. He goes tomorrow with Jacob Steiner to visit the Upper Places.[26] Brother Weigand was there in his place.

2 F Brother Thomas was not there today. Rauschenberger swept chimneys. Weigand was there in Ernst's place, and all the rest were there also.

3 Sa All the Brethren from yesterday were there. John Bruce also helped today. The waterwheel was brought into the raceway [wheelpit] today, and the pumping was stopped. For 22 days we pumped day and night!

5 The Brethren Schober, Franz Seifart, and Jacob Schlung together with Schenke, Gerstberger, and Joh. Knecht came as masons and helpers today. But since it rained in the morning, they had to take a holiday. Christensen, Rauschenberger, and Peter Krist worked on the mill; and Just, Holter, and Andrew hauled away dirt.

6 Tu Everyone was there today except Just, who only worked in the morning.

7 W All 12 worked the entire day today. The carpenters worked yesterday and today on the raceway.

8 Th All were working again. Since Kliest's wife was very sick, he asked me to give work to his boy. So the boy was here yesterday and today and helped Just and Holter dig away the dam. Today I bought the first flaxseed at 6/- per bushel with warranty.[27] It was only 2 bushels.

9 F Schober, Franz [Seifart], and their helpers were working at the cattle yard today. Also David Blum was with Just and Holter.

10 Sa Peter Krist was not there in the afternoon. Schober and his men were still at the cattle yard. The rest were all present.

12 M Scaffolding poles were set up. Schober and his men did not come until after breakfast. Just and Holter hauled away the dams; Schindler helped.

13 Tu All worked except Just, and also on the

14 W all were there except Just, who was absent.

15 Th Everyone from yesterday was at work again.

16 F Today was like yesterday. The missing scaffolding poles were installed, and the scaffold was finished. Just was here yesterday and today.

17 Sa Everyone from yesterday was at work again today.

19 M Because it rained yesterday, last night, and also this morning, no masonry work was done at the oil mill. Schenke went with Franz [Seifart] and Knecht to finish the bakery. The carpenters continued, however. Schober was at home; Gerstberger was sick.

20 Tu Only Schober and his men were at work today. Schenke finished at the bakery.

21 W Everyone was there again, but Andrew only ½ day.

22 Th Everyone was there again, the same as yesterday. Gerstberger returned. The carpenters have been working every day. They started to tear down the [old] dam today. They continued with this on the

23 F and 24 Sa and started to construct the new dam.[28] The masons were all there.

24 Sa The carpenters worked on the [new] dam yesterday and today. Brother Weigand has worked there with Brother Just both days.

26 M They worked on the [new] dam again with Christensen, Rauschenberger, Peter Krist, Weigand, Weisser, and Just present. Brother Gold from Nazareth increased the number of masons.

27 Tu Christensen was at the sawmill with Rauschenberger and Peter Krist. They squared the shaft at the sawmill. The second scaffold was set up today.

28 W All the carpenters were up [here] and worked on the shafts.

29 Th Nobody worked on the building except Schober, Gold, and the 2 helpers.
30 F All the masons, their helpers, and the carpenters were at work. Weigand and Weisser dug out the head race.
31 Sa Everyone who worked yesterday was there today.

SEPTEMBER 1765

2 M and 3 Tu The sluice gate was completed. Weigand and Weisser helped to dig. All the masons and carpenters were there.
4 W The same persons were working.
5 Th Today was like yesterday; the same with the carpenters and masons. The third scaffold was made today.
6 F Rauschenberger swept chimneys today. The beams[29] were installed. Gold went home during the afternoon for his celebration.[30]
7 Sa Rauschenberger, Schober, Gold, and Andrew the Black were not at work. Installation of the beams was completed.
9 M Schober with his Franz [Seifart]. Jacob [Schlung] was at his house.
10 Tu Same.
11 W All the masons and carpenters were there.
12 Th The raceway was installed.
13 F and 14 Sa All the masons and carpenters were there.
16 M All the masons and carpenters worked. The fourth scaffold was built. The remaining part of the new dam was completed but could not be installed yet. The walls for the race trench were started.
17 Tu The above work continued.
18 W The dam[31] was breached, and the water ran through the open sluice gate for the first time.
19 Th All the masons and carpenters worked, also Just, Weigand, and Weisser. The remaining part of the [new] dam was installed.
20 F and 21 Sa Worked again on the [new] dam. Schenke made the dry wall downstairs at the west end of the entrance.[33] I was in Philadelphia this week. The masons and carpenters continued to work.
30 M Nielson and Demuth came from Christiansbrunn to repair the fulling mill under Christensen's direction and to lay pipes from the [water] tower to the stable and the tavern. By Saturday the masons completed the wall to the second floor level.[34] So when the carpenters are finished next week, the [roof beams] can be erected.

57. Exact details of the new oil mill dam built across Monocacy Creek in 1765 have not survived. It apparently had a wooden frame which was built in sections and then installed in the creekbed, probably by loading with stones (see August 22-24, September 16-18 entries). The oil mill dam shown in this 1870 photograph may bear some resemblance to the one constructed a century earlier.[32]

OCTOBER 1765

7 M They pointed the walls inside by the waterwheels.
8 Tu In the morning they continued pointing in the tanbark mill. They started to erect [the roof beams] in the afternoon and on
9 W completed the job. Praise the Lord that everything went so well and without harm.
10 Th They began to wall up the gable-ends and the rafters. They finished the gable-end on the waterworks side on
11 F Gold went home at 4 o'clock.
14 M The carpenters put on the roof over the entrance. The masons put up the central fire wall and started to point the completed gable-end. Brother Richter and his Knaus also helped today. Kunz and Willis moved out of the stable.
15 Tu The carpenters started to lath. The masons worked on the central gable-end.
16 W Since it rained in the morning and the masons could not work, they started tearing down the baking oven. Schenke, Gold, Franz [Seifart], Knecht, and Jacob [Schlauch] worked at this. All were doing masonry work again in the afternoon, and Steph. Nicolaus worked on the stove.
17 Th Today after lunch they started to roof and continued during the afternoon, since the weather was too damp in the morning.
18 F The masons completed the chimney and the last gable-end today. They started to point.
19 Sa The weather was too foul and damp, so they could not put the white pine shingles on the roof. The masons pointed inside. At noon the helpers left; Jacob Schlauch will not return since only 2 are needed now.
21 M It was nice weather. The masons were pointing and the carpenters roofing. Steph. Nicolaus and Henr. Schmid worked on the road.
22 Tu and [23] W Roofing and pointing. Steph. Nicolaus and Henr. Schmid worked on the road and on the baking oven.
24 and 25 Again on the road to the oil mill and
26 Sa on road work for the township. The masons were still pointing outside. The roof was completed in the morning and the flag placed on it.
28 M The stand[35] for the waterwheel was built, and Friedemann started to make the stairs.
29 Tu The masons completely finished on the outside. The carpenters started work on the waterwheel. Schober and Franz [Seifart] left today and will work occasionally around town.
30, 31 and

NOVEMBER 1765

1 they continued working on the waterwheel and the stairs. The masons pointed inside and half

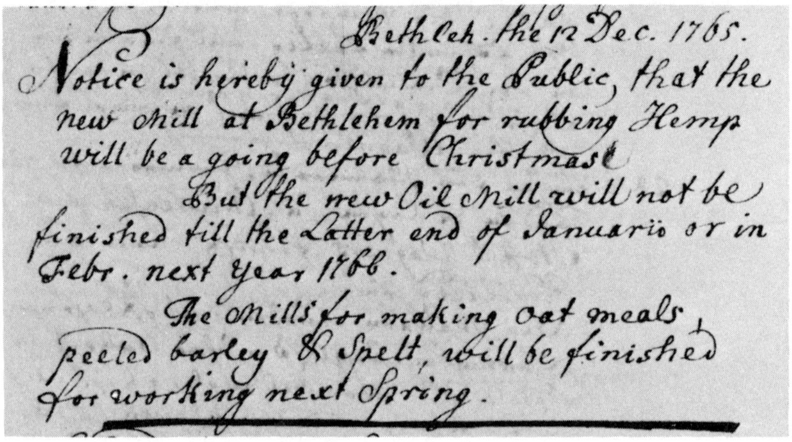
Bethleh. the 12 Dec. 1765.
Notice is hereby given to the Public, that the new Mill at Bethlehem for rubbing Hemp will be a going before Christmast
But the new Oil Mill will not be finished till the Latter end of Januario or in Febr. next year 1766.
The Mills for making oat meals, peeled barley & Spelt, will be finished for working next Spring.

58. The diary entry for December 12, 1765 quotes the announcement posted for the opening of the new oil mill at Bethlehem.

of the small entry hall was plastered.

4 M 3000 boards were selected for the floor of the oil mill.

5 Tu The 3 masons Schenke, Gerstberger, and Andrew, their helpers, and also Brother Gold were still pointing inside. The carpenters worked on the waterwheel and the stairs.

6 W Brother Gold completed his current work here and returned to Nazareth this morning. He worked 60¾ days for 3/- with free board, which comes to 4/6 per day.

7 Th The masons finished pointing inside.

8 F Foundations were built for the stamping troughs in the tanbark mill. The entry hall stairs, which the Brethren Friedemann and Peter Krist had been working on, were put up. Brother Gerstberger plastered the small room. Schenke went to Christiansbrunn this afternoon. The first rim for the waterwheel including the paddles was completed and removed from the stand.

11 M Brother Gerstberger put up the wall in front of the small room at the oil mill. Schenke stayed up at the bakery. The carpenters placed the second rim for the waterwheel on the stand; but since it began to rain, they started to lath by the waterwheels and to finish the dam. Also the boards for the floor were planed and fitted.

12 Tu Continued with it.

13 W This was a holiday.

14 Th The carpenters were laying the floor, and the masons were up here in the old bakery.

15 F and 16 Sa The carpenters were still laying the floor. The masons were in the old bakery until noon Saturday.

18 M Schenke, Gerstberger, and Andrew worked on the small room and built foundations for the [stamping] troughs in the oilseed mill. The first small stairway was installed today, and they worked boarding up by the waterwheels. Christensen attached the second rim on the second waterwheel.

19 Tu The second rim of the waterwheel was completed. Planks were put on the sides of the raceway, and the window boards were installed in the small room. Started to plaster in the oilseed mill.

20 W The posts under the crossbeams in the tanbark mill were installed. Schenke was still plastering, and Gerstberger cleaned up the small room at the oil mill.

21 Th The [iron] bands were driven onto the shafts and the gudgeons inserted.

22 F The large shaft was pulled into place. The sills for the tanbark mill [stamping] troughs were laid.

23 Sa The other two shafts were also brought across. The arms [spokes] of the new waterwheel were fitted in[to the shaft], and the two rims were fastened on. The masons started to plaster in the tanbark mill.

25 M The second waterwheel was almost completed today. The masons were still plastering in the tanbark mill.

26 Tu The new waterwheel was wedged and completed except for tarring. Two of the millstones were brought up.

27 W The other 2 millstones were also brought up. The masons finished plastering in the tanbark mill and started to finish up their remaining work in the oil mill. Christensen bored on a pump for Stoffel Wagner this afternoon.

28 Th The joiners and carpenters started to lay the floor again. Christensen was still working on Stoffel Wagner's pump and also on Horsfeld's. The masons completely finished plastering and concluded their work at the oil mill for now. The only thing left to do is to whitewash the small room.

29 F and 30 Sa The carpenters continued laying the floor.

DECEMBER 1765

2 M and 3 Tu[36] The groat mill shaft, which also lifts the oilseed stampers, was put up.

4 W The [stamping] troughs and all the lumber belonging to the oil mill were hauled down with a sled. During the remaining days of this week they worked on the tanbark, hemp, and leather fulling mill in order to finish it before Christmas.

9 M The door for the tanbark mill was hung and finished today. The arches in front of and behind the waterwheel[s] were closed off with boards to keep out the frost.

10 Tu The work in the tanbark mill continued all week.

12 The following advertisement was posted:

Bethleh. the 12 Dec. 1765.

Notice is hereby given to the Public, that the new Mill at Bethlehem for rubbing Hemp[37] will be a going before Christmas. But the new Oil Mill will not be finished till the Latter end of Januario or in Febr. next year 1766. The Mills for making oat meals, peeled barley & Spelt, will be finished for working next Spring.[38]

14 Sa The troughs in the tanbark mill and a post for the leather fulling mill were installed today.

16 M, 17 Tu and 18 W Work on the tanbark and leather fulling mill continued. The small shaft was in-

stalled and the cams driven in.

21 Sa The tanbark and hemp mills were sufficiently complete that the hemp stampers could be started up at 4 o'clock. This was the first trial.

23 M and 24 Tu The sluice gates in the [new] dam and the grating in front of the waterwheels were constructed. On the latter day the dam was finished and the grating installed.

27 F Georg Rothrock was the first from the neighborhood to bring hemp for stamping so he got 50 pounds [stamped] free. Schowalder and another man were also there, and they all got to take their hemp home.

28 Sa Mich. Puz[39] brought some millet to be stamped.

JANUARY 1766

2, 3 and 4 They worked inside on the oilseed mill and up in Christensen's workshop on the cogwheels. So they continued.

16 Th The uppermost cogwheel was hung.

18 Sa The iron plates were installed down inside the stamping mortars.[40]

20 M They started to work on the spur gear and to install the troughs in the oilseed mill. They set up the prefabricated framework on both sides.

26 From today until February 1 they worked on the large spur gear and on the troughs and related equipment in the oilseed mill.

FEBRUARY 1766

3 From this day on, still continued the above work. On the 5th the [roasting] pan[41] was embedded; and on the 4th, the day before, the spur gear was hung. This work continued the rest of the week.

10 This week the cogs were driven in, the base for the oil trough was laid, and similar work such as installing the small linseed mill.[42] Rauschenberger started to cut tanbark on the 10th.

18 On Tuesday the first flaxseed was crushed for pressing oil.

19 W The first trial oil pressing took place in the evening.

20 Th Flaxseed was crushed again. Since there was too little water, we could not make oil today.

21 F Full scale oil pressing began.

24 M Busy crushing and stamping flaxseed. Pressed 11 gallons and 1 quart [of linseed oil] in the afternoon.

25 Tu This continued until Saturday, March 1, when we had pressed 42 bushels [of flaxseed].

MARCH 1766

3 M The tanbark mill, which had remained idle until now, was made ready to run today. We also started to stamp barley for Boeckel today. The second floor [directly] above the waterwheels was laid this week.

10 M Christensen and David Kunz wanted to go to the water tower to lay new pipes for a new water line to the inn and the stable. But since it snowed, nothing came of it. The tanbark mill and the oilseed mill were both running at the same time this morning. Brother Boeckel's flaxseed was pressed, and he also brought today some that he had gotten from Geitner. It was worked on during the week. Brother Christensen started to hammer the pieces of sheet metal[43] for the groat mill. David Kunz worked on the cogwheel for the hulling mill. Friedemann was in Christiansbrunn this week.

17 M On Tuesday the 18th they finished with Boeckel's flaxseed. Christensen and David [Kunz] worked on the cogwheel for the hulling mill. The clapper mill was finished.

19 W Rauschenberger swept chimneys, worked 1 day stamping tanbark, and the remaining days in the oilseed mill. Peter Krist was in the tanbark mill the last 4 days.

24 to 29th [no entry]

25 Tu No work was done.

27 Th Only a half day.

APRIL 1766

1 Tu[44] Started again. Christensen went to Friedensthal with Schenke to inspect the dilapidated mill today.

3 Th Christensen and Kunz started to prepare the pipes for laying at the water tower and continued on Friday the 4th and Saturday the 5th. This Saturday we completely finished the oil pressing. From 110 bushels of flaxseed we obtained 155 gallons of oil.

7 M Since it rained today, Christensen and Kunz worked in the shop. Peter Krist and Schindler made the floor in the tanbark mill. Rauschenberger made sassafras and laurel oil for Otto.[45]

8 Tu Schindler and Peter [Krist] worked in the tanbark mill. Christensen and Kunz on the second cogwheel.

9 W Christensen and David Kunz started to empty out the water tower and put in the pipes. Peter Krist also helped in the afternoon.

10 Th Christensen, Kunz, and Peter Krist were at the water tower again.

11 F and 12 Sa The same 3 Brethren still working at the water tower. Wagenseil has hauled earth and stones at the oil mill for 4 days. Three pounds of tallow were used at the water tower. Gerstberger constructed the wall on the raceway at the oil mill on Friday and Saturday.

14 M Christensen, David [Kunz], and Peter Krist were laying pipes. Gerstberger and Wagenseil were making stairs at the oil mill.

15 Tu Since it rained today, everyone worked in the oil mill. In the afternoon the Lehigh started to rise. At 6 o'clock it was 48" and at 7 o'clock 57". But at 9 o'clock it was 73".

16 W At 6 o'clock in the morning the Lehigh was 126". All the lumber at the oil mill was lying every which way in the water. The stamping trough in the oilseed mill and the tanbark trough were under water, which rose above the first step of the stairs in the entry hall. They had all they could do today bringing up the lumber and organizing it.

17 Th At 6 o'clock in the morning the water was 90", at 7 o'clock 88". Peter Krist was at Kornmann's, so Weigand took his place with Rauschenberger to clean up at the oil mill. Friedemann worked on the second cogwheel, and Christensen and David Kunz on Dettmer's privy trough today.

18 F Rauschenberger, Schindler, and Wagenseil organized the lumber at the oil mill. Christensen, David [Kunz], and Peter [Krist] were laying pipes.

19 Sa Same as yesterday.

21, 22, 23, 24, 25, 26 [no entry]

26 From the 21st to 26th, the entire week that is, Brother Christensen worked with David [Kunz] and Peter [Krist] at the [water] tower laying pipes, except on Tuesday morning the 22nd, when it snowed. Christensen has also bored pipes for 1½ days, so 2 days must be deducted for him.

Rauschenberger, Wagenseil, and Friedemann were at the mill all the time. Lyhn and Bieninger hauled rubble for landfill for 3 days.

28 M Rauschenberger was in the tanbark mill the entire week up until Saturday. Christensen, David [Kunz], and Peter [Krist] were at the tower laying pipes. Lyhn and Bieninger hauled rubble. Christensen bored.

29 Tu Same as yesterday. Christensen bored [pipes] today; with 1½ days before, this makes 3½ days of boring. Rauschenberger in the tanbark mill.

30 W Christensen, David [Kunz], and Peter [Krist] were laying their own pipes. Rauschenberger, 4 days in the tanbark mill.

MAY 1766

3 Th For this week things were as follows:

Christensen	boring and laying pipes		6 days
Kunz	with him		3½ days
	at the tower		2½ days
Peter Krist	laying pipes		3 days
	at the tower		2½ days
	dam installation		½ day
Holter			1 day
Rauschenberger		tanbark mill	4 days
		oil mill	1½ days
Lyhn & Bieninger	hauling dirt		6 days

[Cost of piping work:]

Laid	324'	3 elbows, separate @ 1/-	3.-
2 risers	22'	2 risers @ 2/-	4.-
Sum	346' @ 3d		

Note: The risers by the inn 10', the others 12'.

£50 per year is 19/3 per week.[46]

5 M Christensen, David Kunz, and Peter [Krist] went fishing. Rauschenberger and Friedemann were in the oil mill.

59. The "Oil Mill Diary" is part of the collection of Moravian manuscripts and imprints at the Moravian Archives in Bethlehem, Pennsylvania. Its quarter leather binding with brown paste paper covers is typical of eighteenth century account books. The title is lettered on a white paper cartouche pasted on the front cover.

6 Tu Christensen, Kunz, and Peter [Krist] planed the risers and installed them.

7 W Christensen, Kunz, and Peter [Krist] went fishing. Rauschenberger and Friedemann in the oil mill.

8 Th Ascension Day, a holiday.

9 F Christensen, Kunz, Peter [Krist], and Rauschenberger went fishing. Friedemann in the oil mill.

10 Sa The above four went fishing. Friedemann in the oil mill.

12 M Christensen, David Kunz, and Peter Krist, ¾ day at Lange's, etc.

13 Tu Fishing, and Rauschenberger.

14 W Christensen, David [Kunz], and Peter [Krist] at Lange's, etc.

15 Fishing, and Rauschenberger.

16 On the pipes again.

17 Fishing, and Rauschenberger.

18 Su Whitsunday.

19 M Whitmonday.

20 Tu Rauschenberger, tanbark mill. Peter [Krist] and David [Kunz], mill construction. Christensen, water pipes.

21 W Same.

22 Th Rauschenberger, Peter [Krist], and David [Kunz], mill construction. Christensen at Payn's.

23 F Rauschenberger to Emmaus in the afternoon.

21-24 Friedemann, Christensen's commode.

24 Sa Rauschenberger, Emmaus. This week:

Peter [Krist] and David [Kunz], 1½ days on the water box.[47]

Christensen, 2 days selecting lumber and fitting planks for the water box.

Peter [Krist] and David [Kunz], 3½ days on mill construction.

26 M Christensen, David [Kunz], and Peter [Krist] on the water trough. Rauschenberger in the oil mill. Friedemann on Christensen's commode.

27 Tu Christensen, David [Kunz], and Peter [Krist] on the water trough. Rauschenberger in the oil mill. Friedemann on Christensen's commode.

28 W Christensen, David [Kunz], and Peter [Krist] on the water trough. Friedemann on the garden gate. Rauschenberger, oil mill.

29 Th Christensen alone on the water trough. David [Kunz] and Rauschenberger stacking boards. Peter [Krist] with his brother Melchior in the sickroom. Friedemann on Christensen's commode.

30 F and 31 Sa Nobody worked. Synod.

JUNE 1766

2 M David Kunz, Rauschenberger, and Peter Krist laid floor in the tanbark mill.

3 Tu David Kunz, Rauschenberger, and Peter Krist laid floor in the tanbark mill.

4 W David Kunz, Rauschenberger, and Peter Krist laid floor.

5 Th David [Kunz] (entire day), Rauschenberger, and Peter [Krist] (¾ day) laid floor. Christensen on pipe work until today.

6 F Everyone on the oil mill construction.

7 Sa Same.

9 M Rauschenberger, David [Kunz], Peter [Krist], and Christensen on the apple press.

10 Tu Same.

11 W Same.

12 Th Rauschenberger mowed. The others on the press.

13 F Rauschenberger making hay. Others, as above.

14 Sa Rauschenberger making hay. Others, as above.

16 M Rauschenberger making hay. Others, as above. Severe thunderstorms and downpour.

17 Tu Rauschenberger, tanbark and oilseed mills. Oth-

ers, as above.

18 W Rauschenberger, tanbark and oilseed mills. Others, as above; the [apple] press was completed. Even more severe thunderstorms and downpour.

19 Th
20 F Rauschenberger, tanbark and oilseed mills. The rest, oil mill contruction.
21 Sa Wagenseil cut tanbark until today.

23 M Christensen, David Kunz, Peter Krist, and Rauschenberger, all on laying floors and mill construction.

24 Tu All the above on oil mill construction until
28 Sa This week:

Peter Krist	oil mill construction	6 days
Rauschenberger	[oil mill construction]	3 days
	in the tanbark mill,	3 days
	stamped 68 mortarloads	
Wagenseil	in the garden all week	

30 M Christensen, Rauschenberger, David [Kunz], and Peter [Krist] were on mill construction. The 2 cogwheels with the lantern pinions[48] were hoisted.

JULY 1766

1 Tu All the above Brethren were again working on mill construction.

2, 3, 4, 5 All were working on construction in the mill.
This week:

Peter Krist	6 days	paid @ 3/6	entered
Rauschenberger	5 days	paid @ 3/3	
chimneys	1 day		

7 M Everyone worked at the oil mill.

8 Tu I traveled to New York. Giers went to Lititz. Rauschenberger in the kitchen. Peter Krist to Boeckel.

16 W I returned from New York.

17 H Rauschenberger returned to work [at the mill] and continued

18 F and 19 Sa
From July 5-20:

Rauschenberger	only 4 days	12.-	paid
Peter Krist	5 [days]	15.-	paid

21 M Christensen, David [Kunz], and Rauschenberger on the oil mill construction.

22 Tu Christensen, David [Kunz], and Rauschenberger on the oil mill construction.

23 W Christensen, David [Kunz], Rauschenberger, and Peter Krist, same. Hauled 2 flatboats of stones.

24 Th All four worked.

25 F Again all four worked.

26 Sa David [Kunz] and Christensen worked. Peter [Krist] and Rauschenberger went fishing. So this week:

Rauschenberger	5 days
Peter Krist	3 days

28 M Christensen, David [Kunz], Peter [Krist], and Rauschenberger.

29 Tu, 30 W, 31 Th

AUGUST 1766

1, 2 All worked until Saturday. This week:

Rauschenberger	5 [days]			
Peter Krist	6			
Rauchenberger	oil mill	9 days	27.-	
	tanbark mill	1 day	3.-	
			30.-	paid
Peter Krist	oil mill	9 days	27.-	paid

4 M, 5 Tu, 6 W, 7 Th, 8 F, 9 Sa
This week: Rauchenberger only 1 day
All the others 6 days

11 M All.

12 Tu Peter Krist at the brick kiln.

13 W Peter Krist at the brick kiln. Rauschenberger sick ¼ day.

14 Th Rauschenberger sick. David [Kunz] and Peter [Krist] at Danielsen's.

15 F Rauschenberger sick. David [Kunz] and Peter [Krist] at Danielsen's.

16 Sa Rauschenberger sick. The others in the mill. Thomas finished the doors.

Peter Krist	12 days	36.-	paid
Rauschenberger	4½ days	13.6	

18 M Christensen in Friedensthal. David [Kunz] laid floor in the mill. Rauschenberger sick. Thomas started to work in the mill. Peter Krist mowed our meadow with Schindler.

19 Tu Peter Krist stamped tanbark, 21 mortarloads. Rauschenberger sick. Christensen and David [Kunz] in the mill.

20 W David [Kunz] and Peter [Krist] on the pump for the brandy distillery. Christensen in the mill.

21 Th David [Kunz] and Peter [Krist] worked on the pump.

22 F David [Kunz] and Peter [Krist] worked on the pump.

23 Sa David [Kunz], Peter [Krist], and Christensen in the mill.

25 M Christensen, David [Kunz], and Peter [Krist] went to Friedensthal, where they remained until

27 W when they returned home.

28 Th The above three worked at the mill.

29 F Choir festival.[49]

30 Sa The above three worked on a pump at Jungmann's former soap boiling works.
From the 18th to the 30th:

Christensen in Friedensthal		4 days
David Kunz in Friedensthal		3 days
Rauschenberger did not work but owes		
Behringer for pressing		
2 bushels of flaxseed	6.0	
Kunckler for tobacco	2.6	
Otto owes for one bushel of flour	2.0[50]	
Peter Krist, mowing meadow		1 day
stamping tanbark		1 "
pumpmaking		3 days
Friedensthal		3 "
oil mill construction		3 "
sum	paid	11 days

SEPTEMBER 1766

1 M David [Kunz] and Christensen on Sangerhausen's pump the entire morning. The afternoon in the oil mill. Rauschenberger stamped oak galls for Weiss.

2 Tu Rauschenberger stamped millet.

3 W He stamped millet.

4 Th He stamped tanbark.

5 F Rode out for Borhek. Christensen and David [Kunz], ½ day on the pump by the brick kiln.

5-6 Masons in the Monocacy.

6 Sa Rauschenberger, David [Kunz], Christensen, all at the mill.

1 pump at the sawmill brandy distillery	
23' long @ 2/-	£ 2.6.-
1 pump at the brick kiln	
11' long @ 2/-	£ 1.2.-
Fulling mill, gudgeons wedged in	1.-

8 M The masons, Gerstberger, Steph. [Nicolaus], Christensen, Kunz, and Rauschenberger at the mill.

9 Tu Rain. David [Kunz], Christensen, and Rauschenberger at the mill.

10 W David [Kunz], Christensen, and Rauschenberger in the mill.

11 Th, 12 F, 13 Sa Same. Made 18 mortarloads of tanbark in 1 day.
From the 13th to 27th David Kunz and Rauschenberger worked as follows:

Rauschenberger	12 days mill construction	36.-	paid
David Kunz	the same		
Sold 4 quarts of oil		6.0	
1 bushel millet stamped		3.0	
1 bushel flour sold		2.0	
1½ bushels flaxseed stamped		7.6	
		18.6	entered
Against it debit			
Bought 8 bushels of flaxseed		32.-	entered

OCTOBER 1766

From September 29 to October 4, Rauschenberger and David [Kunz] with Christensen worked on the alarm device[51] in the groat mill.
From October 5 to 11 the three Brethren worked on the dryer.[52]
From September 29 until October 11, 2 weeks altogether, Rauschenberger spent five days stamping tanbark (86 mortarloads) and worked on the mill the remaining 7 days.

Stamped tobacco @ ½d per pound	1.6	entered
Sold 2 gallons 1 pint of oil	12.9	
Ground 1 bushel barley	3.-	
4 quarts groats	2.-	
1 container for	3.9	
Cleaned 1 bushel	-.6	
	23.6	
Bought 6½ bushels flaxseed @ 4/-	26.-	
Rauschenberger for 12 days @ 3/-	36.-	paid

Christensen and Kunz started to make the water troughs on the 13th.
13 Christensen with Kornmann in the fulling mill. 1½ days @ 4/- 6.0
Kunz worked on the pigsty for 1½ days.
14 Both started to make the water troughs, until the 18th.
18 41 bushels of linseed left, 20 bushels pressed.
From the 12th to the 18th:
Christensen, water trough, 4½ days @ 4/- 18.-
David [Kunz] 4½ days @ 3/6 15.9
20 M Christensen at Schowalder's buying barley. Kunz with Schindler on the fence.
21 Tu Christensen made the grist chutes in the mill and coarse-ground the maize. David [Kunz] helped Schindler with the fence.
22 W Christensen and Schindler worked in the oil mill because it rained.
23 Th Christensen and David Kunz worked on the water trough.
24 F It rained. David Kunz worked up in the mill and Christensen too.
25 Sa Both on the water trough.
Christensen & David [Kunz]
oil mill construction 1 day
water trough 2 days 15.-
Rauchenberger tanbark 1½ days @ 3/- 4.6
oil mill 10½ days @ 2/6 26.3
27 M Christensen and David [Kunz] worked on the water troughs. Rauchenberger, oil mill.
28 Tu and 29 W Christensen and David [Kunz] worked on the water trough. Rauschenberger in the oil mill.
31 F Same.

NOVEMBER 1766

1 Sa The same, except Christensen ½ day less on the water trough. So:

Christensen, this week		6½ days
David Kunz	6 days	
Christensen, last week		4 days
David Kunz	4½ days	
Sum	10½ days	10 days[53]

3 M, 4 Tu Christensen and David [Kunz] on the water box. Rauschenberger and Steph. Nicolaus, oil mill, also 2 days last week.
W and Th Same.
8 F and Sa Again.
Christensen
water trough 4½ days pipe boring 1½ days
David Kunz 4½ 1½
This week Rauschenberger and Steph. Nicolaus pressed oil the entire week.
10 M David [Kunz] and Christensen, pump work.
11 Tu Same.
12 W David [Kunz] and Christensen, water trough.
13 Th Both in the mill. Christensen ground oats.
14 F ½ day on the water trough.
15 Sa David [Kunz] and Christensen on the water trough.
This week: Christensen and David [Kunz]
water trough 2½ days, pump 2 [days]
22 From the 17th to 22nd inclusive:
Christensen water trough 4 days
oil mill 2
David [Kunz] water trough 4¼ days
oil mill 1¾ days
Paid Rauschenberger 30/- for 14 days.
24, 25, 26 Christensen made groats and grist, and David [Kunz] worked on the oil trough and press.
27 Christensen on Kraus's pipes, and David [Kunz] and Rauschenberger pressed oil.
28, 29 Christensen on Knecht's pump. Christensen and David [Kunz], ½ day each on the water trough. 15/- to Rauschenberger. David [Kunz], 2 days pressing oil.

DECEMBER 1766

1 M, 2 Tu Christensen on the pump. David [Kunz] in the oilseed mill.
3 W It was cold and snowed. Christensen worked.
4 Th Christensen on the pump.
5 F In the mill on the pump handle.
6 Sa Christensen, water trough entire day. Rauschenberger 15/-.
8 Monday. David [Kunz] and Rauschenberger made oil. Christensen on the pump handle and similar work.
9 Tu Same.
10 W Christensen made oat meal. David [Kunz], oilseed mill.
11 Th Christensen on the bridge and hulling spelt. David [Kunz], oilseed mill.
12 F Christensen installed Knecht's pump. David [Kunz], oilseed mill.
13 Sa Same.
15 M Rauchenberger, hemp mill. David [Kunz] and Christensen on the millstone casing for the groat mill.
16 Tu Rauchenberger, hemp mill. David [Kunz] and Christensen on the casing ½ day and in the afternoon made Oberlin's block.
17 W Rauchenberger and David [Kunz] pressed oil. Christensen fixed the groat [mill] casing.
18 Th Rauschenberger and David [Kunz] pressed oil. Christensen on the casing for the groat mill.
19 F Rauschenberger and David [Kunz] pressed oil. Christensen worked on the groat [mill] casing and finished it.
20 Sa Christensen made groats. David [Kunz], the stairway.

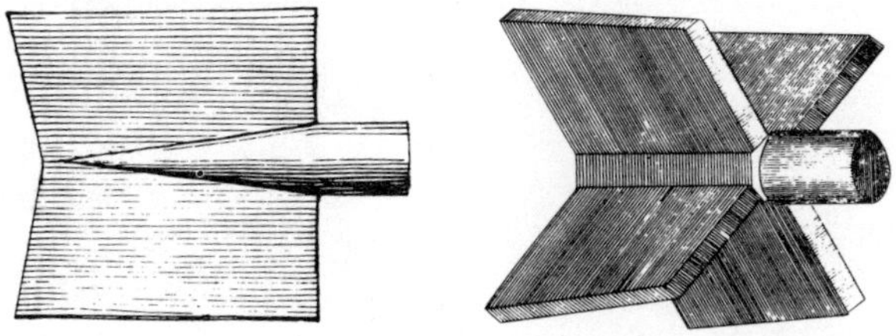

TWO WING GUDGEON   FOUR WING GUDGEON

60. Gudgeons were iron rods fixed into the ends of revolving wooden shafts to provide pins which could be seated in stationary bearings. Wings on the rod prevented it from turning independently.[56]

Christensen	3 days	groat mill
	3 days	oil mill construction, [millstone] casing
David [Kunz]	3 days	oilseed mill
	3 days	stairway

22 M Christensen made groats. Rauschenberger, hemp. David [Kunz] and Schindler, the stairs.
23 Tu Advertised barley on December 15. Christensen buying barley. Rauschenberger, oilseed mill. David [Kunz], stairway.
24 Christensen buying barley. Rauschenberger, oilseed mill. David [Kunz], stairway.
25, 26 First and second Christmas days. 27 Saturday.
29 M, 30 Tu, 31 W Rauschenberger and David [Kunz] in the oilseed mill. Christensen, miscellaneous.

JANUARY 1767

2 F Trouble.[54]
3 Sa Rauschenberger and David [Kunz], oilseed mill.
5 M Water was 90" high. They cleaned barley.
6 Tu Cleaned barley because the mill could not run.
7 W Again nothing could be done.
8 Th David Kunz tore down Astrom's house. Christensen and Rauschenberger, oil mill, but could not do anything.
9 F David [Kunz] and Rauschenberger pressed oil. Christensen went to Funck's on the Little Lehigh [Creek].
10 Sa David [Kunz] and Rauchenberger pressed oil.
12 M, 13 Tu, 14 W, 15[55] Th Christensen, David [Kunz], and Rauschenberger made groats.

Popellwelt	1 pipe for his washing room	
	1 crossarm for his sluice gate rod	2/-

19 M Christensen on Kornmann's pump. Rauschenberger and David [Kunz], pressing oil.
20 Christensen on the mill[stone] casing. David [Kunz] and Rauschenberger, [pressing] oil.
21 Same.
22, 23, 24 Rauschenberger and David [Kunz], [pressing] oil. Christensen, the last 2 days, oat meal.
26 M, 27, 28 Christensen worked on the spelt mill. David [Kunz] and Rauschenberger, oilseed mill.
29, 30 and 31 Christensen worked on the fanning machine.[57] David [Kunz] and Rauschenberger, pressing oil.

FEBRUARY 1767

2, 3, 4, 5, 6 and 7 David [Kunz] and Rauschenberger pressed oil. Christensen completed the fanning machine.
9 to 14 David [Kunz] and Rauschenberger, oilseed mill. Christensen started Funck's mill this week.
21 This week Christensen made oat meal for 2 days. David [Kunz] and Rauschenberger, oilseed mill.
28 The same this week. The oilseed mill will be idle for some time.

MARCH 1767

5 Christensen at the waterworks; gudgeons, 3/4 day, new plugs, 1 day.
7 Made groats this week and [built] the locker for the millstone casing.
Oats, 45 bushels gone. Barley, 177 bushels gone.
11 Wedged in the gudgeons at Stoll's sawmill.
14 This entire week and 2 days last week, worked day and night making groats from 100 bushels of barley.

APRIL 1767

27 Christensen, Leimbach, and Rauschenberger were at the Little Lehigh to start Funck's mill. Returned on July 1, so they worked 57 days up there on it.

Christensen, from February 9 to July 1, 20 weeks 4 days	
Leimbach, March 15 to July 1, 15 weeks 3 days @ 3/- per day	£13.19.-
Rauschenberger, April 27 to July 1, 9 weeks 3 days	7. 2.6
Leimbach, board from March 15 to April 27, 6 weeks 1 day @ 3/6	1. 1.6
Christensen, board from February 9 to July 1, 20 weeks 4 days @ 1/-	7. 7.-[58]

JUNE 1767

15 David Kunz started to make groats from 100 bushels.

JULY 1767

1 W The Brethren Christensen, Leimbach, and Rauschenberger came back home.
2 Th Christensen made a new gudgeon bearing for the oil mill. Leimbach and Rauschenberger sharpened their axes in the morning. In the afternoon they and David [Kunz] hauled stones with the flatboat.
3 F Rauschenberger, David [Kunz], and Leimbach were again hauling stones for the drying kiln by the oil mill.
4 Sa They all went for bleeding, so it was a holiday and only a 5 day work week.
6 M Christensen, Rauschenberger, and Leimbach started to work on the flatboat. Thomas on the new furniture for the children's home.
20 They finished working on the flatboat. They had worked on it for 13 days.

Christensen @ 5/-	£3. 2.6
Leimbach @ 3/6	2. 3.9
Rauschenberger @ 3/6	2. 3.9
	£7.10.-

21 Tu Started to work on the sluice gates at the flour mill dam, but it rained all day and they could not work.
22 W They started again.
23 Th The entire [day].
24 F The entire day.
25 Sa Worked until 4 o'clock when the fire engine was tested.
27 M The entire day.
28 Tu, 29 W, 30 Th, 31 F and

AUGUST 1767

1 Sa 9¾ or 10 days. Note: Accounts settled with Leimbach and Rauschenberger to this point.
3 M The carpenters were supposed to start work at the joiners' shop. They made pegs in the morning and went to the apple mill in the afternoon.
4 Tu Same.
5 W Same.
6 Th Same. In the afternoon John Bruce and David [Kunz] went to the Monocacy to take out gravel.
7 F Same. David [Kunz] and John Bruce at the Monocacy. Christensen and his men, 5 days at the apple mill.
8 Sa They went fishing.
10 M The wooden shed was started.
11 Tu, 12 W, 13 Th [no entry]
14 F Rain in the afternoon.
15 Sa Working on the shed again. Christensen, from Wednesday noon, Thursday, Friday, Saturday, 3½ days at the saw mill. Leimbach, ½ day.
17 M, 18 Tu, 19 W, 20 Th, 21 F, 22 Sa On the wooden shed. Christensen went to Lititz on Monday and returned on Friday evening. On Saturday he went to Christiansbrunn.
24 M, 25 Tu Finished at noon. In the afternoon Christensen, Leimbach, and Rauschenberger started on the apple mill again.
26 W All 3 on the apple mill again.
27 Th Same.
28 F Same.
29 Holiday.
31 M Leimbach and Rauschenberger on the apple mill.

SEPTEMBER 1767

1 Tu, 2 W, 3 Th, 4 F, 5 Sa Christensen was in his shop the entire week.
7 M Married couples festival. Christensen at the flour mill and lovefeast. Leimbach and Rauschenberger building fence.
8 Tu All 3 at the flour mill.
12 From August 31 to September 12:
Christensen, 11 days at the flour mill.
Rauschenberger & Leimbach, 1½ days at Worbas'.
Rauschenberger & Leimbach, 2 days working at Funck's mill.
Settled accounts to date with Rauschenberger and Leimbach, the latter owing his board from August 2 on.
Note: From August 2 to September 12 they have worked 34 days, namely

13½	days at the shed	
1	day at the garden fence	
14	days at the apple mill	
1½	days at the flour mill	5½
1	day at the Widows' House	
3	days at Funck's mill	
34	days	

14 to 19 In Philadelphia.
28 Gerstberger and David Kunz started on the oat drying kiln and continued until October 20. Bruce, 1 day road work, and on October 3 he made firewood. Thomas worked until then for the children's home.

OCTOBER 1767

2 Rauschenberger and Leimbach worked on Funck's mill for 18 days. Leimbach charged for 10 weeks 4 days board, through the 7th.
7 Rauschenberger credit for 21 days £3.3.-
debit for cash 4.-.-[59]
8 Funck's wagon picked up the wheels, also Rauschenberger, Leimbach, and Christensen.
Simon Heller's pump, 9' long @ 2/6.
Beer brewer on the new town pump, 14', 30/-.
Joh. Klein for his pump

40' long @ 2/6	£5. -.-
Brass, 18 pounds @ 4/-	3.12.-
	£8.12.-

Thomas has worked on the move of the children's home from July 6 to September 28, which is 12 weeks but he was gone 2 weeks. That leaves 10 weeks @ 18/- per week which is £9.-.-. The Building account is to be debited for this and the Joiners' Shop account credited.
13 Jacob Herr and Merck started masonry work at the oat drying kiln.
23 Schenke, Gerstberger, and Hübsch on the oat drying kiln.

NOVEMBER 1767

7 Christensen and his men returned home from Funck's, having worked there from October 7 to November 7. To this morning that makes 3 weeks 6½ days or 27½ days.
9 M Started pressing oil; 18 days so far. Christensen to Heller's. Rauschenberger and Leimbach cleaned up in the morning and made wooden pegs in the afternoon. Jacob Herr, Merck, and Kunz made tanbark.
Jacob Herr and Merck worked on the oat drying kiln for 19 days.
9, 10, 11 & 12 Rauschenberger and Leimbach worked here.
13 Was a holiday.
14 At Heller's for the first time.
16 M They went to Heller's again.
27 Boeckel 30 bushels of linseed meal
Angel 1 bushel of hemp seed

DECEMBER 1767

9 They finished at Heller's and came home. They spent 22 days there but were only paid for 20 by Heller.
10 During the morning the water boxes were winterized against freezing. At the store during the afternoon.
11 Again on the water boxes.
12 Again until noon, and at Boeckel's during the afternoon.
14 M At Boeckel's. Hemp stamping during the afternoon.
2 days on the water boxes
½ day in the store
1 day at Boeckel's
½ day stamping hemp
15 Tu Yesterday afternoon they started to work again in the room with the gear works for the hemp roller mill, and finished early in February 1768, namely

FEBRUARY 1768

5 at noon.
Note: They worked on the hemp roller mill from November 9 to noon February 5, which is 73½ days
During that time they were

at Christoph Heller's	22 days	
on the water boxes	2 days	
in the store	½ day	
at Boeckel's	1 day	
must be subtracted		25½ days
Remaining for the hemp mill		48 days

"OIL MILL ACCOUNT BOOK"

As financial supervisor for construction of the 1765 oil mill in Bethlehem, John Arbo kept a detailed record of the project's expenses from 1764 through 1768. His "Oil Mill Account Book"[60] provides us with considerable information about how the mill was built and what types of equipment were installed. Details such as the brass bearings for the power shafts, the compartment built around the tanbark stampers, and the chestnut and pine shingles on the roof are only found here. We also learn from its pages how many quarts of rum and kegs of beer were purchased for the construction workers and included as part of the cost of the building!

The entries found in the "Oil Mill Account Book" also appear in the ledgers of the Single Brethren's Choir, although some pages are missing.[61] The ledger version is somewhat less organized, suggesting that Arbo copied his "Oil Mill Account Book" from it.

Still a third version of the "Oil Mill Account Book" exists. After the building was completed, probably in 1768, Arbo decided the building had been costed too low:

> It was a mistake to set the cost of the building at £1140.13.3 in the first bill. Some things were forgotten; others were not estimated at their true value. Be that as it may, no private party could have built the building for under £1200. The Choir house has given extraordinary help that cannot be put into a bill and for which a private party would have had to pay. Thus, I have charged nothing for my travel and supervision, which would not have been exactly cheap. The same is true for the oat dryer and for the hemp mill.[5]

Arbo then wrote an entire "Revised Oil Mill Account Book," repricing some expenses, adding others that he thought had been forgotten, and changing the dates of some entries. His new cost estimate was £51.12.5 higher than before.

We present here an English translation of the "Oil Mill Account Book" (original version) in its entirety. The book is divided into two sections. The first part is a chronological journal of construction expenses ending with separate compilations for the oat dryer and the hemp roller mill, which were built after the main building was completed. The second part is a summary of all the construction costs by category: masons' wages, ironwork, lumber, stones, etc.

EXPENSES[62]

DECEMBER 1764 — £ . s . d

31	To Brother Schenke for quarrying stones in five different locations to see where the best stones can be obtained	1.12. -
	To Brother Kliest for 4 pair of ice spurs	12. -

JANUARY 1765

19 To the following Brethren for quarrying stones on Astrom's hill[14] and hauling them across [the Lehigh River] with the hand-sleds @ 3/- per day

				£ . s . d
Just	10½	days	1.11.6	
Schenke	11		1.13.-	
Gerstberger	10½		1.11.6	
Schober	10¾		1.12.6	
Franz Seifart	8½		1. 5.6	
Rauschenberger	7½		1. 2.6	
Wm. Volk	6		18.-	
Straus	6½		19.6	
Hornig	11		1.13.-	
Hübsch	11		1.13.-	
Claus Cölln	6½		19.6	
Möhring	3½		10.6	
Danielsen & Jorde	5		15.-	
Schindler	2		6.-	
John Bruce	6		18.-	
James Hall	2½		7.6	
Rum and bread			1.15.6	19.12. -

			£ . s . d
	To Straus for cutting down 6 trees		3. -
24	To the Diaconate for taking over the ironwork and the wooden tools[63]		49. 1. 4
26	To the masons for 53 sledloads of stones from the old lime kiln @ 1¼ perches[64] per sled, which makes 66¼ perches @ 22d		6.1.5½
	To the Diaconate for 11 days of work by Christensen in Christiansbrunn on lumber for the oil mill		2. 4. -
28	To Joh. Becker for lumber as per bill		3.10. 6
	To the pine-wood shop for 2 barrels		12. 6
31	To Just for 5 days quarrying stones		15. -
	To Georg Schlosser for 1 barrel of tar		
	Cost	16.-	
	Freight	9.-	1. 5. -
	To the smithy for various things for quarrying and hauling stones, namely		
	4 rings on the sled runners	4.-	
	4 stone drills	1.4	
	1 stone pick sharpened	-.6	
	1 crowbar and 1 pick repaired	1.5	
	4 rails for sleds	13.-	
	1 stone pick sharpened	-.6	
	1 crowbar sharpened	-.9	
	Same	-.9	
	1 stone pick sharpened	-.3	1. 2. 6

FEBRUARY 1765

7	To Brother Boeckel for hauling lumber and stones	32.14. 6

61. The "Oil Mill Account Book" is also part of the large collection of early Moravian manuscripts and imprints held by the Moravian Archives in Bethlehem, Pennsylvania. Its simple leather and paste paper binding is ornamented with a white paper cartouche containing the title.

NOTE	£ . s . d
Last summer various barges full of stones that had been destined for something else were fished out of the Lehigh since they were good solid wall stones and nice face stones. By mistake the drivers hauled them to the oil mill with the other stones. Since these stones have caused so much trouble and work and are extraordinarily good for the foundation of the oil mill (although I would have much preferred that they had remained where they were), I have credited the Building & Repair account and debited the amount again here @ 8/- per full sled	2. -. -
	121.5.9½
The pump that had been included in the Oil Mill Inventory[65] is subtracted from this, since it belongs to a man named Funck on the Lehigh. He has received it and has paid me	(1.10.-)
	119.15.9½

MARCH 1765

11	For hauling stones across the ice of the Lehigh in January, and for same and also quarrying stones on the Monocacy in February and the beginning of March				
	Just	5½ days	17.6		
	Hübsch	5	15.-		
	Hornig	3	9.-		
	John Bruce	5 days	15.-		
	Rauschenberger	2½	7.6		
	Schober	2½	7.6		
	Schenke	5½	16.6		
	Gerstberger	4	12.-		
	Seifart	4	12.-		
	Möhring	2½	7.6		
	Jorde & Danielsen	1½	4.6		
	Peter Krist	2	6.-		
	Straus	2½	7.6		
	Wm. Volk	2	6.-		
	Schindler	1	3.-		7. 6. 6
	To Brother Horsfeld for leather for the ice spurs				4. -
	NOTE: To date all the stones hauled to the oil mill cost			39.9.5½	
	Add hauling charges to this			16.-. -	
				55.9.5½	
	Without the stone cartage charge that was part of the February 7th payment to Friederich Boeckel				
	To Anton Schmid to make the rollers[66] in the linseed mill and two sheet metal scrapers				11. 6
	To Lazarus for 1 cartload of planks from the sawmill				1. 6
	To Becker for 2 planks for moving stones				7. 6
22	To Peter Krist, 2 weeks 4 days @ 3/-				2. 6. -
	To Rauschenberger, 3 weeks				2.14. -
	To Wm. Volk, 8½ days making the rafts and shingles				1. 5. -
31	To the farm for 12 cartloads of stones @ 2/4				1. 8. -
	To same for lumber loads				12. -

APRIL 1765

4	To Peter Krist, 4 days squaring logs	12. -
	To Rauschenberger, same	12. -
	To Wm. Volk, still owed for 2 days hauling stones	6. -

			£ . s . d
	To the smithy for sharpening the stone picks and wedges	2.6	
	Repairing the flatboat	8.5	10.11
4	To the joiners' shop for 150' pine boards for the boat landing on the Lehigh for moving stones		12. -
	NOTE: As noted on the previous page, the stones cost	39.9.5½	
	2 planks, from the same page	7.6	
	Wm. Volk, 2 days forgotten	6.-	
	And the above 10/11 and 12/-	1.2.11	
		41.4.10½	
	In addition there is the cartage charge from Boeckel for January and March	28.-.-	
	Sum	69.4.10½	
	If there were 345 perches of stone, then one perch would cost 4/-		
6	Paid to Rauschenberger for 4 days		12. -
	To Peter Krist for same		12. -
13	To Rauschenberger for 4 days		12. -
	To Peter Krist, 5 days		15. -

MAY 1765

4	To Rauschenberger, 13 days	39.-	
	Peter Krist, same	39.-	
18	To Rauschenberger, 7½ days	22.6	
	Peter Krist, 8	24.-	8.15. 6
20	To the joiners' shop for the lumber which had been destined for the wash house		
	3 beams 14' long, 3" thick	42'	
	5 " 12'	60'	
	16 " 17'	272'	
	2 " 28'	56'	
	2 " 12'	24'	
		454' @ 12/-	4. 6. -
	28 rafters, 21' long @ 10/- ; 15 gutter beams, 21', same		8.13. 6
	For chestnut wood for the 1100 shingles		1. 5. -
	To Becker for		
	147' of 4" planks @ 21/-	1.10.10	
	94' of 5" @ 25/-	1. 3. 6	2.14. 4
22	To Friederich Boeckel for hauling lumber from March 31		3.12. -

JUNE 1765

11	To Rauschenberger and Peter Krist from May 20 to June 6		
	Former, 6 days	18.-	
	Latter, 8 days	24.-	2. 2. -
	To the nailsmithy for nails for the lime shed		9.11
11	To Just for tarring the waterwheel		4. -
	To same for repairing the flatboat for carrying lumber and stones		14. -
22	To 4 Brethren for hauling lumber from the sawmill with the flatboat for ½ day		6. -
23	To Freyhube for one cartload of lumber from the sawmill		2. 6

JULY 1765

			£ . s . d
6	To Peter Krist from June 24 to July 6, 12 days @ 3/-		1.16. -
	To Rauschenberger, same		1.16. -
	To Just, 7½ days @ 3/6		1. 6. 3
	To James Hall, 2½ days @ 3/6		8. 9
	To Melchior Krist, 4 days @ 3/6		14. -
	To Bechtel for Weigand, 7 days @ 3/6		1. 4. 6
	To Brother Boeckel for cartage from June as per bill		4. 3. 6
	To Joh. Becker for 275 bushels of lime @ 7½d		8.11.10½
	Sum		189.16.10
	From this we must deduct the wages for the pumps and the pipes, which have been counted in here, namely		
	Mar. 15, 16 3 Brethren, 2 days	(18.-)	
	Mar. 18, 19, 20, 21 2 Brethren, 4 days	(24.-)	
	Apr. 15, 18 2 Brethren	(12.-)	
	Apr. 24 1 Brother, ½ day	(1.6)	
		(2.15.6)	
	Furthermore, it is necessary to transfer from the Oil Mill Inventory into the Mill Construction Inventory		
	2 screws in the cover of the lathe	(4.-)	
	Split beech wood	(12.-)	
	7 pounds new ropes ; 20 pounds old ropes	(8.-)	
	2 pump handles	(5.-)	
	Wood for tools	(3.-.-)	
		(4.9.-)	(7. 4.6)
	Remaining		182.12. 4
18	To Johannes Bonn for lumber and cartage		18. 6. 5
25	To the ferry for Christensen when he was searching for curved wood		1. 4
	To the cartwright shop for shovel, stone pick, and spade		2. 6
	For wheelbarrow	14.-	
	Repairing same	2.6	
	Shovel & spade handles	1.7	18. 1
31	To the smithy		
	1 mattock sharpened	-.6	
	1 stone mallet repaired	1.6	
	2 mattocks sharpened	-.6	
	2 new shovels	11.-	
	2 new spades	13.-	1. 6. 6
	To the kitchen for bread and butter for afternoon break		1. 3. -
	To Richter for one water gutter at the lime shed		5. -
	To the cooper shop for 1 ladle		1. 6
	To same for 2 buckets		3. 4
	For 2 ropes for the boys for pumping		3. 8

AUGUST 1765

	For wages from July 6 to August 5				
	For night pumping			6.16.10	
	Rauschenberger	23 days	@ 3/-	3. 9. -	
	Peter Krist	24 days	@ 3/-	3.12. -	
	Melchior Krist	14½	@ 3/6	2.10. 9	
	Just	16	@ 3/6	2.16. -	
	Andrew	14½	@ 3/-	2. 3. 6	
	Weigand	2	@ 3/-	6. -	
	Jacob Ernst	17	@ 4/-	3. 8. -	
	Sturgius	8	@ 3/6	1. 8. -	
	Joh. Knecht	12	@ 4/-	2. 8. -	28.19. 1
	For 3 barrels of beer to July 15				2. 5. -
17	Rauschenberger	12 days	@ 3/-	1.16.-	
	Peter Krist	12 days		1.16.-	
	Just	8½ days		1. 4.-	
	Andrew	12	@ 3/-	1.16.-	
	Holter	11		1.13.-	
	Knecht	12	@ 4/-	2. 8.-	
	Kliest for Blum	4	@ 2/6	10.-	
	Schober	9½	@ 4/6	2. 2.9	
	Seifart	9½	@ 4/-	1.18.-	

£ . s . d

Jac. Schlauch 9½ @ 4/- 1.18.- 17. 1. 9
To Joh. Klein in Lititz for a stove box 2.17. -
For 169 gallons of beer 3.14.2½
To Brother Führer for
1 rope 8.-
1 " 6.-
1 " 4.-
1 " 6.6 1. 4. 6
27 For 4 millstones @ £4.-.-, totals 16.-.-
Cartage, £3.10.- for each pair 7.-.- 23. -. -
To Johnson for 4 thousand white pine shingles @ £3.10.- 14. -. -
Sum 298.5.2½
31 To Brother Richter for 1 mortar box 3. -
For lime and bricks from July and August as per bill from Becker 14.19. 9
To the nailsmithy as per bill 3.12. 1
To the smithy as per bill 1.11. 2

SEPTEMBER 1765

8 For beer 2.13. 7
16 For wages for 14 days as per construction bill 17.16. -
To the Diaconate for the waterwheel as per bill[67] 46. -. -
To the farm for cartage 8. 6. -
For 1 barrel of rum on August 20 6. 8. -
28 For wages from September 16 to 28
To Rauschenberger 12 days @ 3/- 1.16.-
Peter Krist 12 1.16.-
Franz Seifart 10½ @ 4/- 2. 2.-
Knecht 11½ @ 4/- 2. 6.-
Andrew 11 @ 3/- 1.13.-
Jacob Schlauch 9 @ 4/- 1.16.-
Schober 6 @ 4/6 1. 7.-
To Schenke in partial payment 5.15.-[68]
To Gold in partial payment 1.10.-[68]
Schindler, 6 whole days @ 3/- 18.-
To Just for 6 days 18.-
For stones 12.- 1.10.-
To joiners' shop for Thomas, 4 days of plank fitting at the end of July and beginning of August 18.-
To Schindler for 8 whole days 1. 8.- 24.15. -
To same from July 30 to August 16, ten days @ 2/- 1. -. -
425.9.9½
From this is deducted
For Thomas, which has been previously entered (18.-)
The same for Schindler (1. 8.-) (2. 6.-)
Remaining 423.3.9½

OCTOBER 1765

12 For wages from September 28 to October 12
Rauschenberger 12 days @ 3/- 1.16.-
Peter Krist 12 @ 3/- 1.16.-
Knecht 12 @ 4/- 2. 8.-
Andrew 12 @ 3/- 1.16.-
Weigand 3 @ 3/- 9.-
Schober 7¼ @ 4/6 1.12.7½
Seifart 12 @ 4/- 2. 8.-
Schlauch 10 @ 4/- 2. -.-
Glad 3 @ 2/6 7.6 14.13.1½
Just, hauling stones with his own canoe and equipment, 6 days @ 3/6 1. 1.-
For 34 carts of stones @ 6d 17.-

£ . s . d

For 2 wagons filled with
Lehigh stones 4.-
Helping 1 day at the oil mill 3.- 2. 5. -
Schindler, 10 days @ 2/- 1. -. -
To the cartwright shop for shovel and pick handles 2. 8
14 To the joiners' shop for
135' pine boards 13.6
1 board for the sluice gate 2.- 15. 6
To the nailsmithy as per bill from Römelt 1. 8. 6
To Johannes Bonn for beer as per bill 2.19. 1
For 1 barrel of rum 6. 7. 8
Sum 452.15. 4
29 To the joiners' shop for the remaining lumber that was cut for a wash house
11 pieces, 21' long 131' 13.3
4 ditto, 22', 5" × 6" 88'
8 ditto, 8', 5" × 6" 64'
4 ditto, 9', 6" × 5" 36'
188' @ 12/- 1. 2.6 1.15. 9
To same for 430' of boards for under the cornice and on the gable ends @ 10/- 2. 3. -
Wages, 10 days @ 4/- 2. -. -
To Schenke for a wagonload of stones 2. -
Cartage for the stove box received from Klein in Lititz on August 17 8. -
For wages from October 14 to 27
Rauschenberger 11 days @ 3/- 1.13.-
Peter Krist 12 days @ 3/- 1.16.-
Schober 11½ days @ 4/6 2.11.9
Seifart 12 " @ 4/- 2. 8.-
Jac. Schlauch 6 " @ 4/- 1. 4.-
Andrew 11½ " @ 3/- 1.14.6
Joh. Knecht 11½ " @ 4/- 2. 6.-
Steph. Nicolaus 8 " @ 3/- 1. 4.-
Henr. Schmid 6½ " @ 3/- 19.6 15.16. 9
To Richter and Knaus, 1 day 6. -
31 To Joh. Becker for 700 pine shingles @ 7/- 2. 9. -
For 16 whale oil barrels @ 2/6 2. -. -
To the smithy as per bill 1.18. 6
To the nailsmithy as per bill 10.11. 4
For wages from October 28 to November 9
Totals 9.15.10
To Gold for 60¾ days @ 4/6 13.13.4½ 23.8.4½
To Joh. Becker for
578 bushels of lime @ 7½d 18. 1.3
3300 bricks @ 3/- 4.19.- 23. -. 3

NOVEMBER 1765

19 To the carpenter shop for the following building lumber
1153' boards @ 6/- 3. 9.-
965' " @ 6/6 3. 2.9
687' " @ 7/- 2. 8.-
8.19.9
For various types [of lumber], namely
3" × 3", 3" × 4", 5" × 6",
5" × 7", 6" × 7", 6" × 8",
a total of 1296' @ 12/- 7.16.-
400' same @ 10/- 2. -.-
200' of 2" planks @ 12/- 1.4.-
300' of 3" ditto @ 15/- 2.5.-
528' of 4" ditto @ 20/- 5.5.- 8.14.-
1100' laths @ 2/6 1. 7.6 28.17. 3
1 piece 23' long, 12" × 12" @ 10d per ft. 19. 2
1 piece 14' long, 12" × 14" @ 10d 11.8
2 pieces 22', 12" × 12" @ 9d 16.6
75' boards @ 7/- 6.3 1.14. 5
1 piece 19' long, 12" × 12" @ 9d 14.3
1 piece 19' long, 12" × 8" @ 8d 12.8 1. 6.11

			£ . s . d
	To Becker as per bill		
	220 bushels of lime @ 7½d	6.17. 6	
	2½ " @ 9d	1.10½	
	300 bricks @ 3/-	9. -	7.8.4½
30	For various construction lumber as per bill		50.15. 5
	For scaffolding poles and boards, lime shed, etc.		6. 3. 6
	For 1 piece of lumber 15' long, 12" × 12" wide; 1 piece 15' long, 12" × 8"wide; [both] @ 9d	24.6	
	300' boards	18.-	2. 2. 6
	3 pieces 16' long, 8" × 10" @ 8d		1.12. -
	For wages from November 11 to 23		5. 5. -
	To Schenke for 113½ days work @ 4/6		25.10. 9
	To Gerstberger for 99½ days @ 4/-		19.18. -

DECEMBER 1765

16	For wages from Nov. 25 to Dec. 7		4. 6. -
	To Just for 8 carts of Lehigh flagstones @ 2/-		16. -
	To Kliest for a stove door in the small room		8. -
	For 1 piece of lumber, 12" × 12", 15' long @ 8d		10. -
21	For wages from December 9 to 21, namely		3.19. 6
	Rauschenberger 10 days @ 2/6	1. 5.-	
	Peter Krist 12 days	1.10.-	
	Schenke 4½ @ 3/-	13.6	
	Schindler 5½ @ 2/-	11.-	
		Sum above	
26	For joiners' work as follows		
	24 large window frames @ 8/-	9.12.-	
	3 smaller ones @ 6/-	18.-	
	27 windows		
	18 windows with 12 panes @ 6d	5.8	
	4 leaded windows having 8 panes @ 2/-	16.-	
	3 leaded windows with 3 panes @ 2/-	6.-	
	2 windows in the gable-ends of the entrance wing, for 16 panes @ 6d, 2 door windows, for 22 panes @ 6d	11.-	
		Sum	17.19. -
	5 door frames @ 10/-	2.10.-	
	1 large frame @ 15/-	1.10.-	
		4. -.-	
	3 large doors		
	100' pine wood 6/4" 12.-		
	110' oak wood 9.-		
	6 lbs. nails @ 15d 7.6		
	3 pints linseed oil 3.6		
	1.12.-		
	Wages to make [the doors] @ 15/- 2. 5.-	3.17.-	7.17. -
	To the kitchen for bread at the construction site. From July 31 when they started to dig in the Monocacy to November 9 is 12 weeks; I'll take an average of 2/- [per week], which totals		1. 4. -
	For locks		
	2 large ones	19.6	
	1 same	11.6	
	1 same	10.6	
	1 same	9.6	2.11. -
	To Maintenance account for 1500 old bricks for paving @ 2/-		1.10. -
	For 175' of 4" planks @ 20/-	1.15.-	
	310' boards @ 7/-	1. 2.2	2.17. 2

			£ . s . d
	To Johannes Bonn for		
	30 bushels of lime @ 6d	15.-	
	Cartage	9.-	1. 4. -
31	To Schober for a slate blackboard		4. -
	For wages from December 23 to January 4		3. 3. -
	To the store for the bands for the gable windows and [those] in the small room		7.10
	To the smithy as per bill		11.18.10
	To the nailsmithy as per bill		3.19. 9

JANUARY 1766

10	To the tinsmith shop for 1 lantern		6. -
	To the glazier's shop as per bill		23.18. 8
15	To Brother Boeckel for cartage as per bill		14.15. 6
	To Hübner for installing the stove and the tiles for it		2. 2. 7
	For wages from January 6 to 18		
	Rauschenberger 11 days	1. 7.6	
	Peter Krist 12	1.10.-	2.17. 6

FEBRUARY 1766

1	For wages from January 20 to February 1		
	Rauschenberger 11 days	1. 7.6	
	Peter Krist 12	1.10.-	2.17. 6
	To Tommerop for 4 bearings		5. 2. 3
	To Schaf for weaving 14' of hair matting		6. -
	To Schenke for 1 pair of boots		2. -. -
	To the saddler for the "hairs"		1.18. -
13	To the joiners' shop for the following lumber		
	100' of 5" × 6"	13.-	
	350' boards @ 7/-	1. 4.6	
	125' of 2½" planks @ 14/-	18.8	
	14' of 12" × 12" / 14' of 18" × 18" @ 1/-	1. 8.-	4. 4. 2
14	To Franz Steup for 3 small and 3 large wooden shovels		10. -
	To Just for spinning hair yarn, 5¾ lbs. @ 1/3		7. 6
	To the tinsmithy for		
	½ pint measuring cup	-.6	
	1 pint " "	-.9	
	1 quart " "	1.2	
	1 gallon " "	4.-	
	1 large funnel	2.-	
	1 small "	1.-	

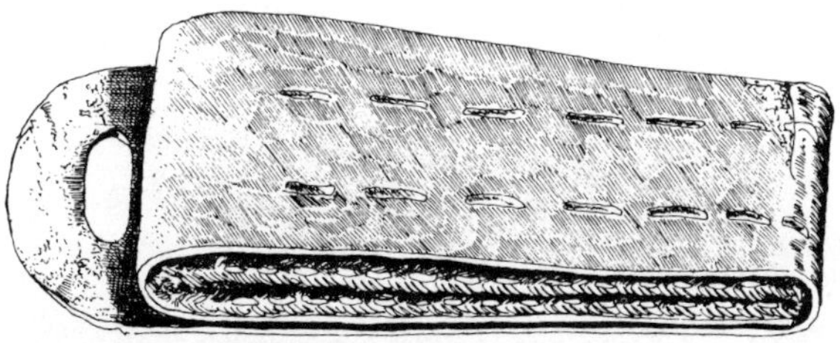

62. The February 1 and 14 entries mention the spinning of hair yarn, the weaving of hair matting, and the making of "hairs" by the saddler. These "hairs" were leather-backed hair mats that were wrapped around the sack of crushed, roasted flaxseed to facilitate oil flow during pressing (see figure 35). This drawing shows one type of "hair" design widely used in central Europe during the late eighteenth century.[69]

			£ . s . d
	2 large kettles @ 18/-	36.-	
	1 piece of sheet metal in hopper	-.5	2. 5.10
22	To Johannes Bonn for 4 bushels of lime @ 9d		3. -
	For 1450' of 5/4" pine boards @ 9/6		6.17. 9
	4 lbs. wool for the oil sacks[70]	6.1	
	For spinning	4.-	
	Knitting 4 sacks	4.-	14. 1
	To Brother Schenke for cleaning up and embedding the [roasting] pan[71]		
	7¾ days @ 2/6	19.4½	
	Gerstberger, 2 [days]	5. -	1.4.4½
	For wages from February 3 to 15		
	Rauschenberger 11 days @ 2/6		1. 7. 6
	Peter Krist 12		1.10. -
		Sum	831.11.4½
28	From this the following amount is to be deducted		
	a) The partial payment to Schenke on September 28 that was listed again on November 30	(5.15.-)	
	Same for Brother Gold	(1.10.-)	
	b) Rauschenberger, for 12 days at the hemp mill @ 2/6	(1.10.-)	
	Same, peeling bark for 6 days	(15.-)	(9.10.-)
		Remaining	822.1.4½
	For 5 business trips on horseback		10. -
	For 1 stove door and the framework under the [iron] water container		10. 6
	4 lbs. of tallow to grease the stampers and gudgeons for the first start		2. -
	To Kliest for 1 stove pipe	10.-	
	4 stove rails	3.4	
	4 springs on the stove frame for the [roasting] pan[71]	1.6	14.10
	4 lbs. of wool for oil sacks[70] @ 1/8	6.8	
	For spinning	4.-	
	Knitting 1 pair	1.-	11. 8

MARCH 1766

10	For 4 advertisements @ 5/-		1. -. -
19	To Geitner for saving 10 bushels of hair[72] for the raceway, which had been forgotten		10. -
	For a one bushel measure		10. -
	For 4 sieves		12. 9
31	To Jungmann for ½ barrel [size] keg		4. -
	To the smithy as per bill from Huber		10.16.10
	To the nailsmithy as per bill from Römelt		2. -. 5
	To the joiners' shop for lumber, namely		
	132' of 3" × 5" @ 12/- per 100'	15.-	
	100' of 10" × 8"	15.-	
	125' of 6" × 6"	15.-	
	114' of 5" × 6"	13.6	
	100' boards	7.6	
	675' same @ 6/-	2. -.6	5. 6. 6
	To Schenke & Gerstberger for plastering up the fire wall under the roof and for quarrying stone at the Monacacy		16. 3
	To the joiners' shop for window moldings		12. -
	To same for the clapper mill	2. -.-	
	To the tinsmithy for sheet metal for it	10.-	2.10. -
	To Mill Construction [account] for wood for making fittings such as cams, shafts, cogs, lantern gears, and the like		3. -. -
	To Anton Schmid for door and window hardware and whatever else was made, as per bill		11.16. 9
	To Brother Christensen for 1 year 10 weeks work @ £50 yearly		59.12. 4

			£ . s . d
	To Friedemann, 160 days @ 3/-		24. -. -
	To David Kunz, 140 days @ 3/-		21. -. -
	For 7½ lbs. weight for the hemp mill @ 3/-		1. 2. 6
	For 1 brass bearing[73] under a gudgeon		1. 2. 6

APRIL 1766

6	To Schindler, 5 days work @ 2/-	10.-	
	To Peter Krist, 2½ days @ 3/-	7.6	17. 6
28	To David Beck, 2 days to clear the pasture		5. -
	To Rauschenberger 3 days @ 3/-	9.-	
	Peter Krist 2½ " @ 3/-	7.6	
	Wagenseil 4 "	12.-	
	Schindler 4 "	8.-	
	Lyhn, 3 days to haul away rubbish	9.-	
	Bieninger, 3 days for same	7.6	2.13. -
30	To store for 1 iron water container	1.-.8	
	1 brush broom	1.2	1. 1.10
	To Gerstberger for the entrance[33] door steps and masonry work on this side of the raceway		12. -
	For 48 lbs. rope @ 1/4 plus freight and commission		3. 6. 8
	To Nicolaus Gauer for		
	654' pine boards @ 7/6	2.8.9	
	300' same @ 9/-	1.7.-	3.15. 9
	For 1 load of sand	1.6	
	To Becker as per bill for 50 bricks @ 4/- per 100	2.-	
	4 bushels lime @ 9d	3.-	6. 6
	To Lyhn for 11 days hauling rubbish for landfill @ 3/6	1.13.-	
	Bieninger, 11 days @ 2/6	1. 7.6	3. -. 6

MAY 1766

14	To the joiners' shop for 2 staircases		
	Wages to make	3.10. -	
	For lumber needed	1.13.11	
	For the box under the clapper mill where the seed falls	8. -	5.11.11
23	For 1200' of boards from Easton @ 7/6	4.10.-	
	Cartage	16.-	
	Expenses	5.-	5.11. -
30	For wages for laying floor		
	Peter Krist, 3½ days @ 3/-	10.6	
	Rauschenberger, 4 days @ 3/-	12.-	1. 2. 6
	Note: David Kunz, who helped with this work, will have his wages figured out when the building is finished.		
31	To the sawmill as per bill from Oerter		3.13.10

JUNE 1766

9	To Borhek for 4 pounds of wool for the oil sacks, which had been forgotten		6. 9
	To the nailsmithy as per bill		1. 3. 8
	For wages from May 24 to June 7		
	Peter Krist, 5¾ days @ 3/-	17.3	
	Rauschenberger, 6	18.-	
	Friedemann, 54 @ 3/-	8. 2.-	9.17. 3

JULY 1766

5	To Rauschenberger, 8 days @ 3/-	1. 4.-	
	Peter Krist, 12 days	1.16.-	3. -. -
	To Oberlin for 14 sacks		3.10. -
20	For wages from July 5 to 20		
	Rauschenberger, 4 days @ 3/-	12.-	
	Peter Krist, 5 days @ 3/-	15.-	1. 7. -

			£ . s . d
23	For 1 rope from Kunckler		7. 6
31	To the nailsmithy		2. 1.10
	To Boeckel for hauling lumber		3. -

AUGUST 1766

2	For wages from July 20 to August 2		
	Rauschenberger, 9 days @ 3/-	27.-	
	Peter Krist, 9 days	27.-	2.14. -
16	For wages from August 4 to 16		
	Peter Krist, 8 days @ 3/-	1. 4.-	
	Rauschenberger, 5 days @ 3/-	15.-	1.19. -
	To Nicolaus Gauer for 910' boards @ 7/6		3. 8. 3
	For wages from August 18 to 30		
	Peter Krist, 3 days	9.-	
	Rauschenberger, 3 [days]	9.-	18. -
31	To Brother Schulze for 2 plans of the oil mill, 1 same of the waterworks. Made for David Nitschmann		1. 7. 6
	To Daniel Kliest as per bill		2.15.10
	To the joiners' shop for the compartment around the stampers in the tanbark mill and for floor laying		7. 3. -

SEPTEMBER 1766

13	To Rauschenberger, 8 days work @ 3/-		1. 4. -
29	For a rope for the stampers		10. -
	To Gerstberger, 2½ days on the other side of the raceway		10. -
	Steph. Nicolaus, 2½ days @ 3/6		8. 9
30	To Rauschenberger, 12 days to build the bridge over the raceway and similar things		1.16. -

OCTOBER 1766

4	To Joh. Becker for lime and lumber as per bill		15. 9
	To Brother Kliest for the stair doors as per bill		6. -
10	To Georg Huber as per bill		21. 7. 2
15	To the store for 2 pair of bands for the dryer		5. -
	To Rauschenberger, 7 days work at the raceway, flaxseed pressing, etc.		1. 1. -
	To the joiners' shop for stair construction and for making a groats box, 16 days @ 2/6		2. -. -
			1076. 2. 5
	From this deduct the overcharge by the joiners' shop, namely on May 14 for the 2 stairways	(1. 2.-)	
	September 3[74] for the compartment around the stampers and floor laying	(1.15.-)	(2.17.-)
			1073. 7. 5
21	3 carts of sand for the wall at the raceway		4. 6
	To the tinsmithy for 3 sheets of metal for the spelt strainer		3. -

NOVEMBER 1766

3	To Kliest for the lock upstairs		2. 6
	To Römelt for nails as per bill		1.12. 3
	To the joiners' shop for 1 table in the small room		1. 6. -
29	To Brother Huber for 2 inch-thick iron plates in the oil trough,[75] 80¼ lbs., as per bill		2.17. -
	For new sheet metal for the clapper mill		18. -
	To Brother Gerstberger for sealing the granary floor		4. -

DECEMBER 1766

			£ . s . d
31	To Christensen for work from April 1 to December 31, 110 days @ 4/-		22. -. -
	To David Kunz, 102 days @ 3/-		15. 6. -
	To Anton Schmid for the window rods		1. 5. 8
			1119. 6. 4

JANUARY 1767

29	For making a winnowing machine to clean barley	19.3	
	For brass wire for it	19.6	1.18. 9
31	For 200' pine boards for 2 boxes	22.-	
	Wages	18.-	
	2 pair of bands for them	3.-	2. 3. -

FEBRUARY 1767

28	To Brother Römelt for nails as per bill		1. -. 1
	To the joiners' shop for 6 staircase beams		13. 6

MARCH 1767

2	To the store for 1 lock for the small room upstairs		2. 6
	To same for 2 pair of bands for the attic windows		6. -
8	To Anton Schmid for 8 window rods		6. 3
10	To the joiners' shop for the following		
	60' boards @ 8/-	4.6	
	26' lumber for 2 window frames @ 15/-	3.9	
	Labor to make @ 3/6	7.-	
	2 window frames @ 2/6	5.-	
	Laying the upper floor and window boards, 14½ days work @ 4/- daily	58.-	3.18. 3
	To the brass foundry for		
	4 bearings, 35 lbs. @ 3/-	5.5.-	
	2 small bearings, 1¼ lbs.	4.-	
	1 small bell	2.6	5.11. 6
	To the glazier's shop for the two roof windows		19. -
	To David Kunz, 13 days work on installation of window hardware, etc.		1.19. -
			1138. 4. 6

MAY 1767

31	To this still belongs		
	The cupboard for the Brethren to store their bread in	14.6	
	For the pea sieve	10.9	
	For a blackboard	2.-	
	A 1 bushel measure, 8/-, and a ½ bushel, 4/6	12.6	
	1 sieve box	9.-	2. 8. 9
			1140.13. 3
	Without tools should be £1152. 7. 3		

OAT DRYING KILN & SHED

18' long	5' in the ground	Main account book C
12' wide	10' above the ground	Folio 270[76]

OCTOBER 1767

10	Wages for stone quarrying and masonry work, 50 perches @ 5/-	12.10. -
	8 beams, 16 spars, 2 frame pieces, 55' of 3" × 4", 600' laths @ 3/-, total	2.19. 7

		£ . s . d
31	To the nailsmithy	5. 4
	To Brother Kliest	10. -
	To haul sand and lime	15.11

NOVEMBER 1767

7	To John Bruce, 3 days	9. -
18	To Boeckel for cartage	2.12. 7
24	For 2 pairs of iron bands	2. 6
	To Römelt for nails	15. 9
	To Kliest for stove door and window rods	11. 3
	For the remaining lumber, door and window frames, boards, beams, laths, etc.	3.14. 3
	To the glazier's shop	3. 6
	For 832 lbs. of iron plates[77] @ 3d	10. 8. -
	To Kliest	6. -
	To the store for 1 lock	3. -

DECEMBER 1767

7	To Becker for lime, bricks, and roof tiles	9.9.11½
	To Gerstberger, 5½ days work inside @ 3/-	16. 6
	For 5 dozen rings[78] @ 6d	2. 6
	To Anton Schmid for the hooks for them	5. -
	To Sangerhausen for the band	2. -
	Lumber for the shed	1.12. -
	To Schindler & Merck for work	15. -
12	To Boeckel for cartage	8. -
	To the saddler's shop for sewing hair cloth	4. 6
	15 yards of hair cloth[78] @ 4/-	3. -. -
	Sum	49.12.1½

			£ . s . d
	For cartage to		
	Boeckel	2.6	
	Conrad Ernst	7.6	
	Christoph Heller	2.6	12. 6
31	To Tommerop for 5 bearings, 28 lbs. @ 3/-		4. 4. -

FEBRUARY 1768

15	To Anton Schmid for 1 pair of bands	6.9	
	To the store for 1 lock	3.9	10. 6
	To Huber for		
	1 bearing, 10¾ lbs. @ 8d	7.2	
	5 rings[80] and 4 gudgeons, 144 ¾ lbs. @ 8d	4.16.6	
	6 irons for the cog wheel	6.-	
	4 bolts installed	2.-	
	4 lantern gear rings[80] made	3. 4.-	
	1 collar around the base, 100 lbs. @ 8d	3. 6.8	
	The axle in the roller stone, 125 lbs.	3.13.-	
	One iron for the rod on the sluice gate	4.9	16. -. 1
	For wages, 50 days		
	Christensen @ 4/-	10.-.-	
	Rauschenberger, 48 days @ 2/6	6.-.-	
	Leimbach @ 2/6	6.-.-	
	Money for drinks	7.6	22. 7. 6
	10 quarts of rum		10. -
	Added for the nails		1. 4. 9
		Sum	70. 3. 7

HEMP ROLLER MILL

NOVEMBER 1767

23	For the roller stone including freight			6.18. -
30	For the following lumber, namely			
	200' of 5" × 6"			
	100' of 3" × 4"			
	300' @ 12/-		1.16.-	
	237' of 8" × 10"			
	96' of 10" × 10"			
	333' @ 24/-		4. -.-	
	730' boards @ 6/-		2. 3.9	
			7.19.9	

DECEMBER 1767

3	300' boards @ 5/-	15.-		
	113' spars @ 4/-	4.6	19.6	
	120' planks @ 20/-	1. 4.-		
	96 cogs @ 1d	8.-		
	30 rungs[79] @ 2d	5.-	1.17.-	

JANUARY 1768

20	To Joh. Schweizer for			
	4 logs for the base			
	1 log for the axle			
	1 log for the main vertical shaft			
	6 logs @ 10/-		3. -.-	
	38' of 10" × 8" @ 24/-	8.-		
	125' planks under the base @ 20/-	25.-	1.13.-	
26	For 590 shingles @ 4/3		1. 5.-	
	For 122' of 6" @ 12/-	14.6		
	280' laths @ 3/-	7.6	1. 2.-	17.16. 3

63. The hemp roller stone purchased for the Bethlehem oil mill on November 23, 1767 probably resembled this one displayed at the Mercer Museum in Doylestown, Pennsylvania. This stone is 29.1 inches (74 cm) high, tapers from 30.7 inches (78 cm) in diameter at the base to 18.5 (47 cm) at the top, and has a hole through the central axis.[81]

SUMMARY OF COSTS[82]

WAGES FOR DIGGING THE FOUNDATION

		£ . s . d
Rauschenberger	40 days @ 3/6	7. -. -
Peter Krist	40 "	7. -. -
James Hall	2½ days	8. 9
Just	55 days	9.14. 3
Weigand	36 days	6. 6. -
Melchior Krist	18½ days	3. 4. 9
Steph. Nicolaus	7½ "	2. 2. -
Henr. Schmid	6½ "	
Feldhausen	2½ "	8. 9
Sturgius	8 "	1. 8. -
Heckedorn	4 "	12. -
Weisser	26 "	4.10. -
Dav. Blum	4 "	
Pell	1 "	3. 6
Starck	2 "	7. -
For beer, rum, butter, and bread		11.17. -
Note: Brother Christensen who helped faithfully is not counted here	Sum	55. 2. -

MASONS AND HELPERS

Schenke	113½ days	@ 4/6	25.10. 9
Gerstberger	99½ "	@ 4/-	19.18. -
Schober	60 "	@ 4/6	13.10. -
Franz Seifart	74 "	4/-	14.15. -
Gold	60¾ "	4/6	13.13.4½
Schlauch	61½ "	4/-	12. 6. -
Knecht	89½ "	4/-	17.18. -
Andrew	93 "	3/-	13.19. -
Jacob Ernst	17 "	4/-	3. 8. -
			134.18.1½
Beer, rum, butter, and bread, from which the carpenters were also supplied			12. 9.10
Schenke, 1 pair of boots			2. -. -
			149.7.11½
Schenke, laying floors, embedding the [roasting] pan, etc., 2 days			5. -
Gerstberger			5. -
Schenke, cleaning up at the construction site and around the entrance, 5¾ days			1.4.4½
To Schenke & Gerstberger, plastering up the firewall under the roof and quarrying stone at the Monocacy			1.16. 3
To Gerstberger for the entrance door steps			12. -
To Gerstberger & Steph. Nicolaus, on raceway			18. 9
			154. 9. 4
Gerstberger, sealing the granary floor			4. -
			154.13. 4

LIME

1503½ bushels @ 7½d	47. 5. 3
Cartage	5.17. -
	53. 2. 3
4 bushels @ 9d	3. -
12 same @ 7d	7. -
	53.12. 3

SAND

152 carts @ 1/6	11. 8. -
3 same	4. 6
	11.12. 6

STONES

	£ . s . d
As per account on page 78, 241 sled loads	69.4.10½
To Brother Just for 6 days hauling stones from Astrom's hill in his own canoe @ 3/6	1. 1. -
To same for 34 carts of stones	17. -
To same for 2 wagonloads of flagstones	4. -
To Schenke for 1 wagonload	2. -
To Just for 8 cartloads of flagstones	16. -
Cartage for above stones	3.14. -
March 31, 1766	75.18.10½
To be deducted because of an error	(1.11.10)
Remainder	74. 7. -

BRICKS

5300 @ 3/-	7.19. -
1500 @ 2/-	1.10. -
Cartage, 17 carts @ 1/6	1. 5. 6
	10.14. 6
50 same	2. -
	10.16. 6

MISCELLANEOUS

For ropes	1. 4.6	
18 kegs @ 2/6	2.12.6	
7 barrels of tar	1. 5.-	
Tarring the waterwheel	4.-	
1 slate blackboard	4.-	
1 lantern	6.-	5.16. -
For 4 lbs. of wool for oil sacks	6.1	
Spinning and knitting	8.-	
4 lbs. of wool	6.8	
Spinning	4.-	
Knitting one pair	1.-	1. 5. 9
To Just for		
5¾ pounds of hair yarn @ 1/3	7.6	
Weaving 14'	6.-	
Saddler's work on it	1.18.-	2.11. 6
Various business trips on horseback		10. -
4 lbs. tallow for lubrication		2. -
For 4 large and 2 small shovels		10. -
For ½ pint measure	-. 6	
1 pint	-. 9	
1 quart	1. 2	
1 gallon	4. -	
1 funnel	2. -	
1 smaller funnel	1. -	
2 large kettles	36. -	
1 sheet metal in hopper	-. 5	2. 5.10
For 4 advertisements		1. -. -
To Geitner for 10 bushels of hair		10. -
1 bushel measure		10. -
4 sieves		12. 9
1 keg for cold-pressed oil		4. -
6 sheets of metal for the clapper mill		10. -
		16. 7.10
To Schulze for 3 building plans		1. 2. 6
1 brush broom		1. 2
1 rope, 48 lbs. @ ¼d plus freight		3. 6. 8
4 lbs. of wool for oil sacks		6. 9
14 sacks		3.10. -
1 rope from Kunckler		7. 6
1 rope from Philadelphia		10. -
		25.11. 6

VARIOUS TYPES OF LUMBER INCLUDING CARTAGE

1 new lantern gear and roller[66] framework	1.10. -
6 logs from inventory	2. 2. -
Cartage	1.16. -

	£ . s . d
6 hickory [logs]	2.18. 6
Cartage	5. 6
	8.12. -
25 logs construction lumber	3.10. 6
2 planks for [hauling] stones	7. 6
241' planks	2.14. 4
Cartage as per bill of January 31	6. 2. 6
From Christiansbrunn as per bill of March 31	18. 6. 5
Cartage for Boeckel as per bill of March 31	2. 4. -
2 logs for pegs, cogs, etc.	6. -
Cartage as per bill of June 30	1.11. 6
150' boards for hauling stones	12. -
To Straus for felling 6 trees	3. -
To the Diaconate, paid for Christensen in Christiansbrunn	2. 4. -
To Lazarus for one load of lumber from the sawmill	1. 6
For floating the wood for the chestnut shingles [down the river]	1. 5. -
Lumber that had been cut for the wash house	4. 6. -
Same	8.13. 6
Repair of the flatboat to transport lumber	6. -
To Freyhube for 2 loads of lumber from the sawmill	2. 6
Ferry money for Christensen when searching for curved wood	1. 4
For 135' pine boards	13. 6
1 board for the sluice gate	2. -
The remaining lumber that had been destined for the wash house	1.15. 9
430' boards for the cornice	2. 3. -
Joiners' work on it	2. -. -
To the joiners' shop for various lumber	28.17. 3
Same	4.17. 9
To Becker as per bill	56.18.11
To the joiners' shop	4. 4. 6
Planks and boards with cartage	3.11. 2
Cartage	14. -
For the old waterwheel with fittings as per bill	37. -. -
To the joiners' shop for lumber	4. 4. 2
1450' of 5/4" pine boards @ 9/6	6.17. 9
To the joiners' shop for lumber	5. 6. 6
To the mill construction shop for wooden fittings	3. -. -
	227.15.10
654' pine boards @ 7/6] 300' same @ 9/-]	3.15. 9
1200' same from Easton	5.11. -
To the sawmill	3.13.10
Cartage	3. -
910' boards @ 7/6	3. 8. 3
Lumber from Becker	8. 9
	241.14. 5
1 table in the small room	1. 8. -

IRONWORK

	£ . s . d
Used, taken over as per inventory	30.14.10
To Anton Schmid for repair of the rollers and two sheets of metal	11. 6
To the nailsmithy as per bill	3.12. 1
To the smithy	1.11. 2
To the nailsmithy	1. 8. 6
To the smithy	1.18. 6
To the nailsmithy	10.11. 4
For 5 locks	2.11. -
To the smithy	11.18.10
To the nailsmithy	3.19. 9
For the bands in the leaded windows	7.10
6 iron rings[80] for the oil mill shaft @ 10/-	3. -. -
2 gudgeons @ 35/-	3.10. -
4 irons with rings[80] for the bark stamper @ 5/-	1. -. -
1 stove pipe	10. -
4 stove rails	3. 4
4 springs on the stove frame for the [roasting] pan	1. 6
1 stove door and framework under the water container	10. 6
The stove	5.16. 1
To Huber according to bill	10.16.10
To Römelt	2. -. 5
To Anton Schmid	11.16. 9
	108.10. 9
One iron water container	1. -. 8
To the nailsmithy	1. 3. 8
To same	2. 1.10
To Dan. Kliest	2.15.10
2 pair of bands for the dryer	5. -
For bands, etc. on the stair door	6. -
To the smithy as per bill	21. 7. 2
	137.11.11

		£ . s . d
For 3 metal sheets for the spelt sieve	3.-	
For the lock upstairs	2.6	
To Römelt for nails	1.12.3	
2 metal plates in the oil trough	2.17.-	
Sheet metal for the clapper mill	18.-	
For the window rods	1. 5.8	
To Römelt for nails	1.-. 1	
For 1 lock	2.10	
Bands for 2 roof windows	6. -	
Window rod	6. 3	8.13. 7
	Sum	146. 5. 6

BRASS BEARINGS

	£ . s . d
Used, taken over as per inventory	3.15. 6
2 bearings, 12 lbs. @ 2/6	1.10. -
4 same from Tommerop	4. 2. -
	9. 7. 6
To Tommerop for bearings	5.11. 6
	14.19. -

SHINGLES

	£ . s . d
1100 chestnut	6. -. 6
4000 white pine @ £3.10.-	14. -. -
700 same @ 7/-	2. 9. -
	22. 9. 6
For 50 cedar shingles	5. 6
	22.15. -

GLAZING WORK

	£ . s . d
Glazing work and painting	23.18. 8
For 2 roof windows	19. 2
	24.17.10

MILLSTONES

		£ . s . d
Millstones	16.-.-	
Cartage	7.-.-	23. -. -

JOINERS' WORK

	£ . s . d
Door and window frames	25.16. -
For the clapper mill	2. -. -
The box underneath it	8. -
Window molding	12. -
2 stairways built and lumber for them	4. 1.11
Compartment around the tanbark stampers and laying floor	5. 8. -
2 stairways constructed, work	2. -. -
	40. 5.11

			£ . s . d
For the boxes			6. 1. 3
			46. 7. 2
1 table in the small room			1. 6. -
			47.13. 2
6 staircase beams			13. 6
			48. 6. 8

FOR VARIOUS WORKMEN'S WAGES

1765			
Rauschenberger	128 days @ 3/-	19. 4.-	
	27 " @ 2/6	3. 7.6	22.11. 6
Peter Krist	135 days @ 3/-	20. 5.-	
	31½ " @ 2/6	3.18.9	24. 3. 9
From January 1 to Febuary 28, 1766			
Rauschenberger	33 days @ 2/6		4. 2. 6
Peter Krist	36 " @ 2/6		4.10. -
Christensen, 1 year 10 weeks @ £50 yearly			59.12. 4
Friedemann	160 days @ 3/-		24. -. -
David Kunz	140 days @ 3/-		21. -. -
Sum to March 31			160. -. 1
April			6.16. -
May			1. 2. 6
June			9.17. 3
July			4. 7. -
August			5.11. -
September			3. -. -
October 21			1. 1. -
		Sum	191.14.10

		£ . s . d
December 31, 1766		
Christensen, 110 days work @ 4/-	22.-.-	
Kunz, 102 days @ 3/-	15.6.-	37. 6. -
		229. -.10
To David Kunz, 13 days work putting hardware on the roof windows and shutters		1.19. -
		230.19.10

RECAPITULATION - May 31, 1767 [83]

Digging foundation	55. 2. -
Masons and apprentices	154.13. 4
Lime	53.12. 3
Sand	11.12. 6
Stones	74. 7. -
Bricks	10.16. 6
Miscellaneous	25.11. 6
Lumber including cartage	240.16. 5
Ironwork	146. 5. 6
Brass	14.19. -
Shingles	22.15. -
Glazing and painting	24.17.10
Millstones	23. -. -
Joiners' work	48. 6. 8
Carpentry work and the like	230.19.10
	1138. 4. 6
Oat drying kiln	49.12.1½
Hemp roller mill	70. 3. 7
[Total building costs]	1258.-.2½

"CALCULATION OF THE MASONRY AT THE OIL MILL"

Reflecting his early training as a mathematician, Arbo sometimes made elaborate calculations of the building materials used in the construction projects that he supervised. The following calculations for the stone masonry work at the 1765 oil mill appear in one of Arbo's notebooks and are presented here in English translation. They provide considerable detail concerning the dimensions and construction of the building.

One immediately notes that these calculations are dated October 31, 1765, long after all the building stone had arrived at the site and thirteen days after the last stone had been laid into place.[84] If these figures were not copied from an earlier reckoning, then they probably were a mathematical after-analysis by John Arbo, an excercise to sharpen his ability to estimate building materials and costs. The fact that he made his calculations in three different ways to check the accuracy of two simplified methods makes the latter interpretation seem more likely.

CALCULATION OF THE MASONRY AT THE OIL MILL[85]

October 31, 1765

Perches[87]

a) The first foundation[88] for the west half of the building, 4' thick

BH = 31½'
EF = 31½'
EB = 17'
FH = 17'
97' long
4' thick
388
2¼' high, times the other
873 - - - - - - - - - - - 35 27/99

b) The first foundation in the second part towards the waterworks, 3' thick

CD = 30' long
AG = 30'
CA = 18'
DG = 18'
96'
3' thick
288
2½ deep
720 - - - - - - - - - - - 29 9/99

Sum for first foundation 64 36/99

c) The second part of the foundation

CD = 30'	BH = 30'
AG = 30'	EF = 30'
CA = 20⅓'	BE = 20⅓'
DG = 20⅓'	HF = 20⅓'
100⅔'	100⅔'
1½' high	3' high
151	302
2½' thick	2½' thick
377½	755
	377½
	1132½ - - 45 75/99

Sum for entire foundation to ground level 110 12/99

d) From ground level to the first floor,[89] i.e. the first story

AB = 66' long
HG = 66'
BH = 26'
EF = 26'
AG = 26'
CD = 26'
236'
2' thick
472
12¾' high
6018

e) From the first floor to above the waterwheel, the one wall

CD[90] = 26' long
2' thick
52
2½' high
130
6018
6148 The first story 248 40/99

The first story with the foundation 358 52/99

Perches

f) The second story

AB = 66'
HG = 66'
BH = 26.8'
EF = 26.8'
AG = 26.8'
212'
1⅔ thick
353⅓
10' high
3533⅓ Second story 142 75/99

g) The 3 gable walls in the roof

30' wide 15' high
7½
225
1½ thick
337½ one gable
3 such
1012½ The 3 gable walls 40 90/99

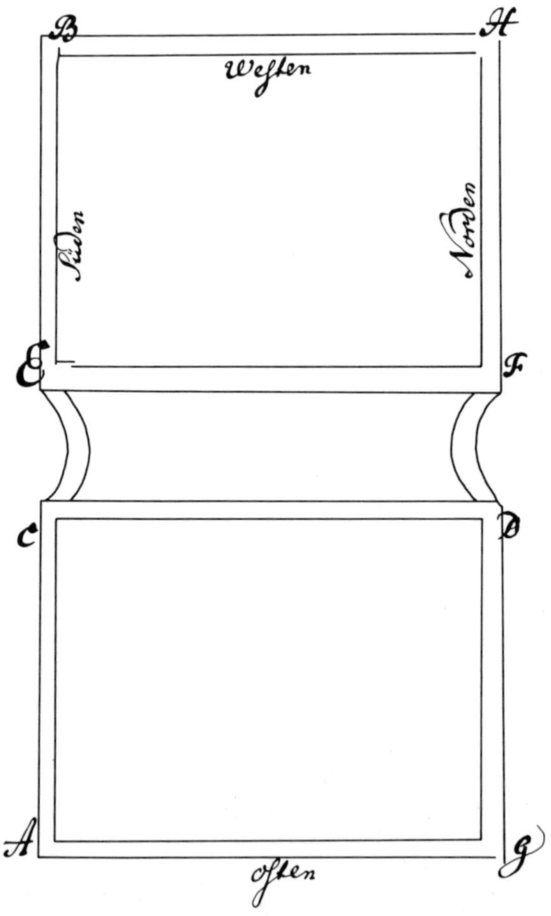

64. Arbo made this drawing designating the various walls in the 1765 oil mill to assist him in calculating the amount of stone needed for the building.[86] *The direction that each wall faced is indicated:* ***AB*** *south,* ***BH*** *west,* ***GH*** *north, and* ***AG*** *east. At the center* ***(CEFD)*** *is the pit for the waterwheels.*

Perches

h) The foundation for the entrance wing

25' long
10' wide
10' wide
45'
3' thick
135
2½' high
337½ the foundation [337½]

i) Up to ground level or second foundation

45' long
2' thick
90
3' high
270 - - - - - - - - - - 270

k) The entire story

45' long
12¾' high
573¾
1½' thick
860⅝ - - - - - - - - - 860

l) The 2 gables in the entrance wing

10' long
5' wide
50
1½' thick
75
2 such
150 - - - - - - - - - - 150

1618 65 37/99

607 56/99

NOTE: Masonry supports for the gudgeon bearings and the 4 troughs; the water walls in front of, behind, and next to the dam; also the upper part of the chimney as high as the rough stones go ——— altogether should amount to 42 43/99

Sum 650

Would at 3/- per perch cost £97.10.-

Rough estimate of how much stone one would need for the oil mill if the walls were two feet thick throughout

Short Side
26'
25' high
650
2' thick
1300
4 such
5200

Gable End
30' long
7½' high
225
2' thick
450
3 such
1350

Long Side
66'
25' high
1650
2' thick
3300
2 such
6600
5200 short side
1350 gable end
13150
÷ 24¾'
[Total] 531⅓ perches

Alternatively

Long Side
66'
2 two times
132

Short Side
104' 4 times 26'
236
25' high
5900
2' thick
11800
Gable 1350
13150
÷ 24¾
[Total] 531⅓ perches as above

NOTE: Did not take the dividing walls or the gudgeon bearing supports into account.

APPENDIX —— CHRISTENSEN AND ARBO MEMOIRS

By custom each member of the eighteenth century Moravian Church was asked at mid-life to write a *Lebenslauf*, a short autobiographical sketch. The text was updated periodically and then completed by the minister at the member's death. This memoir was read at the member's funeral service and preserved in the church records.

The short biographies of Hans Christensen and John Arbo presented in this book have been drawn to a great extent from their funeral memoirs. Naturally we have emphasized their secular accomplishments and their relationships with the Bethlehem oil mill. But to fully comprehend the fervent religious spirit that pervaded eighteenth century Bethlehem and provided the religious motivation behind such commercial efforts as the oil mill, one must read the original memoirs. The intense pietistic emotions of that era are best conveyed by the words of those who lived in it.

To assist those readers who wish to know more about Christensen, Arbo, and their times, we present complete English translations of their memoirs here.

HANS CHRISTOPH CHRISTENSEN [1]

I was born on February 7, 1716 at Hammeloff near Hadersleben. I lost my father when I was three years old, and I was the youngest of four children. During my childhood I was especially impressed that when my mother put me to bed she always prayed so earnestly with us, and particularly the passage "The Blood of Jesus Christ, the Son of God, cleanses us of all sin."

In my sixteenth year I attended the preparation class for Holy Communion and learned much about the Savior and his sufferings, which filled me with happiness. This was especially the case while participating in my first Lord's Supper, when I had the feeling that the Savior was bodily present. Afterwards, however, my love of worldly things returned, and I lost my holy feelings.

However, an uneasiness pursued me until my 22nd year when my elder brother[2] visited me. He was then serving in a royal mill in Hadersleben and had been touched by the Savior. The following day I accompanied him to Hadersleben. On this occasion he had a heart-to-heart talk with me. In a way, I had to agree with him. I fully realized the sinful state I was in, and I could not lose this conviction.

Soon afterwards the King organized a militia in Holstein, and many young people were drafted into it. In order to avoid this, I took work at the mill where my brother was and [where] all were awakened people. They had withdrawn from the world and sought to lead a Christian life. Thus I escaped becoming a soldier because it was a mill which was protected by royal privilege, and nobody could be drafted from it.

During my service in the mill I followed a very strict lifestyle and had many struggles against sin. This lasted until I became acquainted with the [Moravian] Brethren at Stepping; they taught me about the Savior and His merits and the grace freely given through His blood. This consoled and strengthened my heart. I was so carried away by this that it was not too difficult for me to walk two German miles [14.8 km] there and back at night every 14 days just to spend a Sunday with the Brethren in Stepping. Oh, how often afterwards in the Congregation I felt depressed and ashamed when I realized that I was indifferent to the Congregational service. And I asked myself, where is the former zeal, love, and concern that you once had?

Here in Stepping I also heard about Herrnhut and about the Brethren and children of God that were there. Immediately I felt in my heart, that is where you belong also. When I told this to the Brethren, they were helpful and arranged that another Brother and I could accompany a captain's widow as servants. She wanted to go to the Congregation [at Herrnhut] and had a royal pass, which allowed us to travel safely and unimpeded. And so in this company I arrived safe and sound at Herrnhut on May 28, 1744 and to my embarassment was sincerely welcomed by the Brethren.

In September of the same year I was received into the Congregation, and on March 25, 1745 I was allowed to participate in the Lord's Supper with the Congregation. These were two truly festive moments for my heart and helped me in the path of grace. On September 14 of the same year I was received into the Single Brethren's Choir by Brother Johannes, an event which has remained an unforgettable blessing to me.

In December I went with 13 Single Brethren via Niesky to Neusalz, where I remained for five years working all that time as a carpenter to establish the place. During that time I had many trials in spiritual and worldly matters, but my faithful Savior through His inexpressible grace turned everything for the best. At the end when I had persevered, He turned my faith to hold me closer [firmer] to Him.

In September 1750 I was recalled to Herrnhut. In June 1751 I set out, having received a call, to travel from Herrnhut via Zeist to Pennsylvania. On July 4 I sailed on board the "Irene" in the company of the Brethren Nathanael [Seidel], Chr. Seidel, [John Jacob] Schmick, and David Zeisberger; and we arrived safe and sound at Bethlehem in September. I personally had a difficult voyage and was sick almost continually during the crossing.

My first work in this country was in the mill at Friedensthal where I built a new grinding mill. This was the start of my subsequent mill construction work. After this I built the first oil mill along with a tanbark mill and a hemp mill. Then I started on the waterworks, which was not finished at that time because the pipes would not hold.

In 1754 I was sent to Carolina were I built the grist and saw mill near Bethabara,[3] and in the fall of 1756 I returned to Bethlehem again. During this time, in 1755, there was the danger of Indian raids and massacres in the area. Therefore, Bethabara had to be made safe by surrounding it with a palisade. While felling timber I experienced an especially merciful protection of my life and body. A falling tree broke off a large limb from another tree, and this heavy limb hit the spot where I had stood a few moments before. Without any reason I had just moved back a few steps, and the limb just missed me.

In February 1757 I went to Lititz with my three helpers to build a mill there. We completed the first grinding mill on November 11 of that same year, and the entire grist and saw mill was finished in November of the following year. During this time I had the additional duty of supervising the young people who came to visit from the surrounding area.

65. The locations of European towns mentioned in the Christensen and Arbo memoirs are shown on this map together with modern European political boundaries.

After various repair jobs on our mills, I was assigned in 1761 to start building the waterworks again. I took this [assignment] in the name of the Savior and with confidence in Him. He presented me with the necessary intellectual and physical powers to complete the waterworks successfully in June 1762 for the benefit of the Bethlehem Congregation. In 1764 I completely rebuilt the Bethlehem oil mill, which had burned down with all its equipment during the night of November 18, 1763. I also completed the mill in Hope, [New] Jersey between May 1770 and March 1771.

I shall always remember a particular expression of grace protecting and preserving my life, [an incident] in which the entire Bethlehem Congregation took great interest. Just at the end of the year 1768, on December 30, three Brethren and I were on rafts on the swollen Lehigh River. The force of the water ripped the rafts apart, and none of us could help each other. But in that night of danger each one of us was saved from instant drowning by the help of the holy angels.

In spring 1775 I built a new flatboat in Bethlehem, and I suspect that I caught a bad cold in the water. This was the initial cause of my subsequent illness. Later it was so aggravated by my boring pipes in Jersey that I had to spend the fall of 1775 and the following winter of 1776 mostly in the sickroom. Finally on April 14, 1776 I went to Lititz with my two helpers in order to rebuild the burned-down mill.

[Authorship changes.] So far our late Brother's own account.

Here in Lititz, despite great bodily weakness, he supervised and even labored as much as he could. He successfully completed the first grinding mill so that it could run by July 22, and under his supervision the second grinding mill and the hulling mill were also completed and operable before his death. But he finally had to withdraw from everything, and on August 6 he retired to the infirmary for greater care.

We can say something further about his life among us. He was an example of a faithful Brother who was not self-seeking but was concerned about affairs of the Lord and the Congregation. He was a leader and faithful assistant in his Choir and had the spiritual and worldly well-being of his Choir Brethren close to his heart. Since 1770 he was a respected member of the *Aufseher Collegium* in Bethlehem. The customs and rules of the House of God were agreeable and important to him, and he could be very zealous when he knew of any infractions.

By nature he had a somewhat rough and authoritative manner, which sometimes made it difficult for those who had contact with him or worked under his supervision. However, everyone will have to agree that he acted out of a faithful mind and heart. Halfheartedness and apparent unfaithfulness anywhere angered him very much. One can say of him that he lived as a poor sinner [redeemed] by God's suffering and that his heart was in trusting fellowship with his Savior.

His illness was a consumption, from which he died on September 15 in the 61st year of his life, having the blessing of the Congregation and his Choir.

JOHN ARBO[4]

I was born on June 6, 1713 in the parish of Sörup, near Angeln on the Baltic Sea in the district of Flensburg. My parents watched over me carefully in my youth and raised me to be God-fearing according to their belief. In 1745 my mother left this world as a sinner saved through the work of Jesus. Since my childhood I had felt in my heart many manifestations of grace, and already as a child I had been concerned about my salvation. Thus I frequently went alone to pour out my heart before God, and many times I experienced extraordinarily blessed hours that way. However, I was outstanding among my school comrades.

When I was seven years old, I came down with smallpox which affected me so badly that I was incapacitated for all types of work. In my fifteenth year when I went to my first Lord's Supper, I had a special and unforgettable experience. I made a new covenant with God promising to be and remain His. My most blessed and pleasurable times were Advent, Christmas, and Lent; because then there was more singing and preaching about the Savior, and I always enjoyed it. During the same year I was accepted as a helper in the large school that I had been attending, and in my eighteenth year I received my own village school to take care of as schoolmaster.

At the same time, however, I was a servant of sin. My conscience condemned me, and my uneasiness and worry about my salvation greatly increased. In general, I went through many marvelous experiences during those years. In retrospect I see that I could have lost [God's] grace quite often had it not been that a higher power wanted me to be saved and miraculously protected me. Now I was totally immersed in worldly things, but I was always uneasy and could not and dared not do everything that I would have liked to.

In 1737 I received a call to Flensburg to become the Assistant at the German Writing and Arithmetic School of [St.] Mary's Church. This was totally against my previous liking, because I did not like city life at all. Later, however, I realized that this was a part of my being called to the [Moravian] Congregation.

In 1738 several students came home from [the University of] Jena and started services with one of the preachers there. It caused a great excitement in the city. Many people were awakened, some of whom later joined the Congregation. But I did not visit them in the beginning, because I had gotten involved with dissenters and separatists. I was searching for peace and found it nowhere, because the Savior and his merits were still hidden from me.

In 1739 I had a remarkable dream. The Savior appeared to me in the form of a serious man with many bruises and wounds and covered all over with dried blood. He talked to me; and in the forest where I was, he pointed out to me at a distance an old, rotten tree which had neither branches nor leaves. He said, "This is the tree of life. Go to it and climb it. This way you will cross the large, muddy ditch into the land of Canaan to come to the children of God." Quite naively I went to the tree but had not asked how I was to climb it. When I came close to it, I was puzzled about how to climb a tree without branches. A child in white clothing came and placed a ladder on the tree without saying a word to me. So I climbed up; and before I realized it, I was also across the broad, muddy ditch in the land of Canaan. I was not aware how it had happened. Here the children of God received me, welcomed me, and took me along to visit the rest of their friends in that land.

This dream made me think very seriously, and the unsightly, blood-covered man whom I had seen always remained before my eyes. From that hour on, my main desire was to get to know the Savior and to become certain of my salvation by the forgiveness of all my sins through His blood. However, I felt very hampered, my sins depressed me, and I would not and could not yet go to the Savior as a godless sinner. It went so far that many times I was about to despair, and I gave up everything for lost. But still I always looked toward the Savior, cried, prayed, and gave Him no peace, day or night, until He had mercy upon me.

Finally He gave me the grace so that I could let go of everything that had hitherto stood in my way and could throw myself, as I was, on His mercy and grace. Then I had an indescribable feeling of well-being. I felt and believed that the Savior had forgiven me everything and had received and accepted me into His grace. For 14 days I was as if I were love-sick so that I forgot to eat and drink. Many questioned me because of this and asked what was wrong with me. As soon as I lost these sentiments, however, I became deadly melancholic again.

At the same time I became acquainted with some awakened souls who knew something about the Congregation and had visited in Pilgerruh. In 1740 a fellow countryman gave me an account of Closterbergen. I wrote to Abbot Steinmetz asking permission to come, and I received an answer in his own hand granting me permission. In 1741 I happened to see some printed material of the Congregation from which I formed an idea about the Congregation that later proved to be wrong. I often thought about these people and their doctrine, which was quite new but very appealing to me; and it captured my whole mind.

In my catechism lessons for the children I also started to bear witness for the Savior. Sometimes I was too enthusiastic, and I could rightfully have been dismissed. Meanwhile, I was being persecuted a great deal, and several times I was accused at the Consistory as well as at the local Council. A few ministers in the city also preached strongly against those who followed the new belief and especially against me. However, since the Consistory Councilor, the First Mayor, and some other ministers were awakened people, my persecutors were unsuccessful in attaining their goal. But my father, who had to suffer much because of me, declared that he would not recognize me as his son.

The previously mentioned publications of the Congregation and the acquaintance with the Brethren who had visited in Pilgerruh and with those who passed through here on the way from Herrnhut to Copenhagen, all made the Congregation important to me. The urge to join it became stronger and stronger; but since I still had other interests, I was always undecided. This created many difficult struggles within me. Sometimes when I only halfheartedly resolved to go to the Congregation, I felt well. As soon as I changed my mind again, I got restless. Finally I realized this and was greatly perplexed about it.

When I reached the height of my dilemma and did not know how to make up my mind, I went to my room, threw myself on my face, and begged the Savior with bitter tears to clarify in my heart whether this urge to join the Congregation came from Him or if it were only a fantasy. I then got up, took my Bible, and opened it to the words of the prophet Ezekiel, chapter 12, verses 2-4:

> Son of man, thou dwellest in the midst of a rebellious house, which have eyes to see, and see not. They have ears to hear, and hear not; for they are a rebellious house.
>
> Therefore, thou son of man, prepare thy stuff for removing, and remove by day in their sight; and thou shalt remove from thy place to another place in their sight. It may be they will consider, though they be a rebellious house.
>
> Then shalt thou bring forth thy stuff by day in their sight, as stuff for removing; and thou shalt go forth at even in their sight, as they that go forth into captivity.

While reading this I felt as if the Savior were bodily present and that He said those words to me. Therefore I resolved to follow the word of the Lord and go to the Congregation. I immediately wrote for permission to leave, which despite many difficulties I received. This happened at the beginning of 1742.

As soon as it became known that I wanted to leave, temptations came to me from all sides. The pious ministers begged me to stay. The Consistory Councilor made me one offer after another, requesting only that I should remain in the district. If my conscience did not allow me to stay where I was, he offered to take me into his house, take care of me as well as possible, and let me instruct his children. I received offers of very respectable positions from four different places. But every time I remembered what I had read in the "Berlin Sermons"[5] regarding the sixth petition: "When the world offers tremendous glories, a child of God will immediately think 'Do not lead me into temptation'." Thus I could not accept these [offers] to the amazement of everyone.

When I was ready to travel, Brother Carl Fri[e]drich was forcefully arrested by soldiers near Flensburg while he was traveling from Copenhagen to Herrnhut. Since I was active in securing his freedom, I had to postpone my journey for some time. Finally I started it on September 11, 1742 and arrived in Herrnhut on the 24th of the same month.

Instead of being happy and contented as I had expected, I began to feel so badly that I shocked myself. I could love neither the Savior nor the Congregation, and I even felt enmity in me toward the latter. My heart was cold, dead, and without feeling so that I felt myself heavily burdened down. The workers to whom I complained about my distress treated me with more than motherly love, and their kindness to such a man as I considered myself was incomprehensible to me. Meanwhile I had innumerable conversations with the Savior, and I wept quite often before Him.

This continued until December 8th of the same year when I was unexpectedly received into the Congregation. I was likewise showered with a stream of grace, and now I could dearly love the Savior and His people. On January 19, 1743, I had the unforgettable privilege of being a communicant of the martyred body and blood of our Savior with the Congregation.

In October of that same year I stayed for ten days with the Pilgrim Congregation at Gnadek, where I saw the late disciple [Zinzendorf] for the first time. On January 30, 1744 I was accepted for the hourly intercession.[6] In July I received a call to go to Marienborn from where I was soon transferred to the boys school with the Pilgrim Congregation in Lindheim.

From May to October 1745 I was a bookbinder and scribe for the late disciple in Marienborn again. After the departure of the Pilgrim Congregation, I returned to my previous station at Lindheim. On June 4, 1747 I was accepted by the Synod at Herrnhaag as an acolyte, and in August of the following year I moved back with the institution to Marienborn again. What impressed me most during the years 1745, 1746, and 1747 were the blessed Choir celebrations of the Lord's Supper of that period, which I will never forget as long as I breathe.

In October 1749 I was made Director in the Office of Scribes and also acted as Private Secretary to the late Brother Vierorth. In 1751 I moved with a part of the children's school to Hennersdorf and served happily there as well as in Niesky. Then in 1756 I came to Herrnhut as Registrar and Secretary of the Administrative Office.

On February 13, 1757 I had the privilege of being a choir disciple for the first time. On July 5, 1758 I was appointed Assistant to the Warden of the Single Brethren, but for the time being I retained my position at the Administrative Office. On February 17, 1759 the Lord's Supper of the Congregation was especially memorable for me. The dear Savior certainly has a surprisingly faithful and compassionate heart; how much the Holy Ghost labors to teach us. It is indescribable how much Jesus loves us and unbelievable how we try Him.

In May 1759 I completely took over the bookkeeping and accounts of the Single Brethren's Diacony here. On July 16, 1760 I received my call to Pennsylvania, and on February 15, 1761 I was ordained a Deacon of the Brethren's [Moravian] Church. On the 23rd of the same month

I started my journey here [to Pennsylvania], where I arrived on October 25th of the same year.

[Authorship changes.]

So far the late Brother.

We can bear witness to the fact that here too he performed his duties as Warden of his Choir with all loyalty and industry for eleven years. He also carried out the duties given him several years ago as Business Agent for the Missions Directorate with all diligence. By nature he had an active and quick demeanor which often offended one or the other of the Brethren. However, they were soon convinced of his good intentions and that he was only concerned that everything was done according to the will of Christ. His willingness to be of service to everyone also made him popular with outsiders, but his own Choir will miss him the most.

He had been very sick previously in October but had fully recovered. However, on the 4th of this month [December 1772] he caught a serious illness with a high fever. It became worse from day to day so that one could do nothing but wait for his early deliverance. On the evening of the 10th when he was asked whether he would go now to the Savior, he answered, "I am subject to the Savior's will. He may do with me as He pleases." Early on the 11th when some Brethren were visiting him, he said in a weak voice, "I am now going to my rest." This happened soon afterwards with the blessing of his Choir and the Congregation in the 60th year of his life.

NOTES

ABBREVIATIONS

ABU	*Archiv der Brüder-Unität*, Herrnhut, German Democratic Republic
BDT	Bethlehem "Daily Times"
BGT	Bethlehem "Globe-Times"
BPL	Bethlehem Public Library, Bethlehem PA
EMHL	Eleutherian Mills Historical Library, Greenville, Wilmington DE
HBI	Historic Bethlehem Inc., Bethlehem PA
MAB	Moravian Archives, Bethlehem PA
MAS	Moravian Archives, Winston-Salem NC
NYPL	New York Public Library, New York NY
SBG	South Bethlehem "Globe"

FRONTISPIECE

1. Gilman, T., "Model of Bethlehem Oil Mill Built by Stephen Young," photograph, 1980.

INDUSTRY IN EARLY BETHLEHEM

1. Spangenberg, A. G., "An Account of the Manner in which the Protestant Church of the Unitas Fratrum, or United Brethren, Preach the Gospel and carry on their Missions among the Heathen," English translation, London, 1788, 128 pp. Spangenberg was an influential leader in early Bethlehem.
2. The Moravians set up preaching stations in Pennsylvania, Maine, Rhode Island, New York, and New Jersey in the 1740's and established settlements at Winston-Salem NC and Lititz PA in 1753.
3. For further information on the early history of the Moravian Church see: Hamilton, J. T. and K. G. Hamilton, "History of the Moravian Church, the Renewed Unitas Fratrum, 1722-1957," Bethlehem, 1967, 723 pp.
4. Acrelius, I., "A History of New Sweden," English translation by W. M. Reynolds, in "Memoirs of the Historical Society of Pennsylvania," vol. 11, p. 403, 1874.
5. For an interesting discussion of the artist and his prints see Cilella Jr., S. G., "Eighteenth Century Views of Bethlehem" in "Pennsylvania Heritage," vol. 6, pp. 10-14, 1980.
6. Garrison [Jr.], N. "A View of Bethlehem, one of the Brethren's Principal Settlements in Pensylvania [*sic*]," print engraved by T. Noual, London, 1757 (MAB, BPL).
7. Yates, W. R., et al., "Bethlehem of Pennsylvania. The First One Hundred Years," Bethlehem, 1968, pp. 38-39.
8. Erbe, H., "*Bethlehem, Pa. Eine kommunistische Herrnhutter Kolonie des 18. Jahrhunderts*," Stuttgart, 1929, 191 pp. (MAB). Erbe, H., "Bethlehem Pa., A Communistic Herrnhut Colony of the 18th Century," unpublished English translation, Bethlehem, 1959, p. 173 (HBI, MAB). All our page citations refer to the English translation.
9. A detailed description of the workings and the eventual breakdown of the General Economy is given in Gollin, G. L., "Moravians in Two Worlds," New York, 1967, pp. 131-215.
10. Erbe, p. 62.
11. Pownall, T. (L. Mulkearn, ed.), "A Topographical Description of the Dominions of the United States of America," Pittsburgh, 1949, p. 107. Modern punctuation has been added to the passage quoted.
12. Gollin, p. 163. Erbe, pp. 67-68.
13. Erbe, p. 70.
14. The diaries, ledgers, and journals of the Single Brethren record numerous sales of Moravian products to buyers outside Bethlehem. A typical entry reads: "In the afternoon Brother Leschinsky came back from Philadelphia where he sold 3 wagonloads of linseed oil for a good price." "*Chor-Diarium der Ledigen Brüder* [Single Brethren's Choir Diary]," Nov. 1, 1783 (MAB).
15. Erbe, p. 61.
16. The Monocacy Creek rises near the south slope of Blue Mountain and flows 18 miles (29 km) south to join the Lehigh River at Bethlehem. The flow in the lower 6 miles (10 km) of its course was sufficient to power ten mills in 1874, and the heavy clay and limestone underlying its valley easily supported the weights of the dams and mills. See: Beers, D. G., "Atlas of Northampton County, Pennsylvania," Philadelphia, 1874, pp. 13-21, 47; and Miller, B. L., "Topographic and Geologic Atlas of Pennsylvania, No. 206, Allentown Quadrangle," Pennsylvania Dept. of Forests and Waters, 1925, pp. 20-23, 149ff.
17. Murtagh, W. J., "Moravian Architecture and Town Planning," Chapel Hill NC, 1967, pp. 32-34.
18. Anonymous, Map of Bethlehem, [1772] (ABU; photograph at MAB, Rinker films, roll 49). Buildings identified using a key on the original. We date this map 1772 since the 1771 dyehouse is present but the 1773 Single Sisters' House (not shown here) is absent.

19. Murtagh, pp. 22-93.
20. Murtagh, pp. 30-32.
21. "*Haupt-Buch vor die Roth-Gerberey in Bethlehem. 1765* [Ledger for the Bethlehem Tannery. 1765]," ff. 1-24, 173-180, 197-200 (MAB, file box "*Roth-Gerberei*").
22. Annual profits at the tannery averaged £765 for the period 1776-1780 compared to £153 for 1770-1776. "Ledger of the Diaconat at Bethlehem . . . ," E, f. 357; F, ff. 137, 262 (MAB).
23. Ellsworth, L. F., "The American Leather Industry," Chicago, 1969, p. 11.
24. Levering, J. M., "A History of Bethlehem, Pennsylvania 1741-1892," Bethlehem, 1903, p. 236.
25. For detailed studies of the Bethlehem waterworks see: Huetter, K. Z., "The Bethlehem Waterworks," Bethlehem, 1976, 17 pp.; Hindle, B., "The Meaning of the Bethlehem Waterworks," Bethlehem, 1977, 23 pp.; and Rau, R., "Historical Sketch of the Bethlehem Water Works," Bethlehem, 1877, 10 pp.
26. Oberholtzer, E. P., "Philadelphia. A History of the City and its People," Philadelphia, [1911], vol. 1, pp. 406-408.
27. Butterfield, L. H., M. Friedlaender, and M. Kline (eds.), "The Book of Abigail and John. Selected Letters of the Adams Family, 1762-1784," Cambridge MA, 1975, p. 167.
28. Garrison [Jr.], N., and Oerter, "A View of Bethlehem in North America," print engraved by J. Spilsbury, 1784 (MAB).
29. Anonymous, ["Luckenbach Grist Mill"], photograph, c. 1880 (HBI P-4458). We date this photograph c. 1880 because the 1877 north boiler house chimney is present but the 1885 western addition to the mill is absent.
30. Anonymous, "The Old Tannery," photograph, late 1800's (HBI P-1350C).
31. Anonymous, "The Bethlehem Roller Flour Mills," in "1865-1915 Semi-Centennial. The Borough of South Bethlehem, Pennsylvania. Souvenir History," [South Bethlehem], 1915 (BPL). "At Work Again," BDT, Jun. 24, 1869, p. 3. "The Old Bethlehem Mills," BDT, Sep. 27, 1871, p. 1.
32. Anonymous, ["Automobile Junkyard along Monocacy Creek"], photograph, c. 1950 (HBI P-2849).
33. Gilman, T., "Restored 1761 Tannery Building," photograph, 1980 (HBI P-7122).

MORAVIAN OIL MILL COMPLEX 1745-1814

1. Reichel, W. C. (ed.), "Memorials of the Moravian Church," vol. 1, Philadelphia, 1870, pp. 286, 306, 366 (lamp oil). Bloy, C. H., "A History of Printing Ink, Balls, and Rollers," London, 1967, pp. 1-25. Gehret, E. J., and A. G. Keyser, "Flax Processing In Pennsylvania from Seed to Fiber" in "Pennsylvania Folklife," vol. 22, no. 1, p. 22, 1971 (medicine). The Bethlehem oil mill often sold small amounts of linseed oil to the local apothecary, presumably for compounding medicines; see "Ledger of the Diaconat at Bethlehem . . . ," D, f. 328 (MAB).
2. The love feast was a traditional Moravian ceremonial meal, often consisting of rolls and coffee. The "oil cakes" (*Oelkuchen*) mentioned here presumably were cakes deep fried in oil like a doughnut.
3. "Bethlehem Diary," Feb. 12, 18, 1745, pp. 363, 370 (MAB).
4. ["List of Persons, Possessions & Places at 1752"], p. [6] (MAB, file box "Administration. Publica: Taxes 1743-1772").
5. Anonymous, "Bethlehem Tract with all the adjacent Lands," map, [1752] (MAB, map & print collection, drawer I, folder 1, no. 4, copy 134). Since the 1752 oil mill and the 1752 northern extention of the Single Sisters' House are present but the 1754 *Kinderanstalt* and the 1754 waterworks are absent, this map must have been drawn in 1752 or 1753. Because the 1745 oil mill building is still referred to as the *Oel-Prese* and the new dam and raceway for the 1752 oil mill are missing, we believe the map was drawn in 1752 before the 1752 oil mill had been completed.
6. Anonymous (possibly N. Garrison Jr.), "Bethlehem, Pa," watercolor drawing, [1752] (MAB, map & print collection). We date this drawing 1752 because the 1752 apothecary and 1752 northern extention of the Single Sisters' House are present but the 1752 oil mill is absent. The 1745 oil mill was identified by comparison with the map in fig. 9. The second building south of the oil mill appears to have a waterwheel; it is part of the "Tannery" (fig. 9) and probably is the separate tanbark mill mentioned in the 1752 tax listing (note 4).
7. In his memoir (see Appendix) Christensen claims "I built the first oil mill" in 1752. This statement seems odd, since he must have been acquainted with the 1745 oil mill. Could he have considered that the 1745 "oil press" was not a true "mill" because it lacked waterpower? E. Myers has stated that the 1745 oil mill was powered by a horse ("The Culture of Flax in Bethlehem . . . ," BGT, Oct. 2, 1926, p. 6). We have found no data to either confirm or deny this statement.
8. "Ledger of the Diaconat at Bethlehem . . . ," B, ff. 8, 131; C. f. 62 (MAB; ledger B in file box "General Diacony: 2 ledgers, 1743-1748"; ledger C in file box "Gen. Diacony: Ledger of the Diaconat of Bethlehem 1749-1755"). The 1744-1745 entries in ledger B follow an abnormal calendar. The 1744/1745 transition occurs in late March (old style) while the 1745/1746 change occurs on January 1st (new style). For comparison with the 1745 entries in the "Bethlehem Diary" (new style), the starting date of ledger B is taken as Jan. 1, 1745.
9. Ibid., B., f. 50.
10. "Bethlehem Diary," Aug. 17/28, 1747, p. 613; Oct. 17, Nov. 10, 1752, pp. 904, 1033 (MAB).
11. "House Conference, 1750 [-1752]," Oct. 20, 27 1751; Jan. 5, 26, Jun. 14, 1752 (MAB, file box "Single Brn., House Conference: 1750-1752 . . . ").
12. Ibid., Jan. 26, 1752.
13. Unless otherwise indicated, all biographical details about Hans Christensen are taken from his memoir (see Appendix). For other biographies of Christensen see: Jordan, A. F., "Some Early Moravian Builders in America" in "Pennsylvania Folklife," vol. 24, no. 1, pp. 5-8, 1974; Myers, R. E., "Danish Millwright a Genius" in Allentown Sunday "Call-Chronicle," Jun. 5, 1977, p. F-9; Myers, E., "A List of Mills Built by Christiansen [*sic*]," BGT, Nov. 8, 1927, pp. 6, 13.
14. Matthias Christensen (1707-1787). His memoirs survive at the Christiansfeld Moravian Archive in Denmark (P.A. II-R7.10 and P.A. II-R7.18).
15. Jordan, pp. 2-17.
16. "Ledger of the Diaconat at Bethlehem . . . ," D., f. 236; E, f. 190 (MAB).
17. Memoir of Hans Christensen (see Appendix). Sentence order changed for this quotation.
18. The list of mills built for Moravian communities was taken from Christensen's memoir (see Appendix) and the "Oil Mill Diary" and confirmed from other sources. The brevity of Christensen's stay when building mills for private owners implies that he was called in mainly for construction of the complex machinery rather than to put up the building.
19. "Ledger of the Diaconat at Bethlehem . . . ," E, f. 168; f. 190, May 28, 1764 (MAB).

20. Ibid., f. 190, Nov. 15, 1764.
21. "Oil Mill Diary," Mar. 27, May 20, Jun. 24, 1765 (MAB).
22. Ibid., Mar. 26-Sep. 17, 1771.
23. Ibid., Feb. 9-14, Apr. 27, 1767. "*Chor-Diarium der Ledigen Brüder* [Single Brethren's Choir Diary]," Dec. 16, 1767 (MAB).
24. "Single Brethren's Choir Diary," Nov. 7, 1767.
25. "Oil Mill Diary," Oct. 21, 1769.
26. "Single Brethren's Choir Diary," Apr. 10, 1774.
27. Ibid., May 16, 1774.
28. Ibid., Oct. 4, 1774.
29. "*Specification aller unserer Gebäude und Landereyen am 31 May 1758* [Specifications of All Our Buildings and Lands on May 31, 1758]," Bethlehem II, bldg. 12 (MAB, file box "Extracts of Bethl. Accounts 1747-1765. Specification of Buildings & Grounds 1758").
30. "House Conference," Jul. 26, Aug. 23, 1752.
31. Garrison [Jr.], N., "A View of Bethlehem, one of the Brethren's Principal Settlements in Pensylvania [*sic*]," print engraved by T. Noual, London, 1757 (MAB, BPL).
32. Anonymous, "A View of Bethlehem," watercolor drawing, [1755] (MAB, map & print collection, drawer "Bethlehem"). We date this drawing 1755 because the 1754 *Kinderanstalt* is present but the 1756 sawmill is missing.
33. Pownal[l, T.], "A View of Bethle[he]m, the Great Moravian Settlement in the Province of Pennsylvania," print engraved by P. Sandby, London, [1768] (MAB, framed print). This engraving is based upon a sketch made by Pownall when he visited Bethlehem in 1754; see Pownall, T. (L. Mulkearn, ed.), "A Topographical Description of the Dominions of the United States of America," Pittsburgh, 1949, pp. 102, 106.
34. Huetter, K. Z., "The Bethlehem Waterworks," Bethlehem, 1976, pp. 2, 8-9.
35. "Bethlehem Diary," Oct. 17, Nov. 10, 1752, pp. 904, 1033.
36. Rau, R., "The Oil Mill," undated manuscript, p. 136 (MAB, Robert Rau Collection, file box "Papers on Trades & Industry," folder "Oil Mill"). Rau's original source for this statement is unknown to us.
37. "*Gerberey Buch auf alle Monat durchs Jahr 1764 April Monat, sowohl Gerberey als Öhl-Mühle* [Tannery Book for All Months through April 1764, Both Tannery and Oil Mill]," 1762-1765, 182 ff. (MAB, file box "Tannery (*Gerberey & Öhl-Mühle*). Ledger: 1762-1765. Warden: Accounts, 1829").
38. "Extract of the Oeconomy Book from 1753 [to 1762]," ff. 2, 5, 9 (MAB).
39. "House Conference," Jun. 21, 1752. "Ledger of the Diaconat at Bethlehem . . . ," C, f. 175, Dec. 31, 1753. Rapeseed oil was widely used for cooking and lighting during the eighteenth century. It was pressed from the seeds of the rape plant, an oilseed crop grown in northern Europe.
40. Levering, J. M., "A History of Bethlehem, Pennsylvania 1741-1892," Bethlehem, 1903, pp. 287-288.
41. "*Chor-Diarium der Ledigen Brüder*," Nov. 18, 1763.
42. "Bethlehem Diary," Nov. 18, 1763, p. 225.
43. Ibid. Wechquetank was located 22 miles (35 km) northwest of Bethlehem.
44. "Yesterday in the Morning we received News . . . ," anonymous letter, Nov. 18, 1763, 2 pp. (MAB, file box "124. Wechquetank, Pa. 1760-1764," folder "Wechquetank Documents on Indian Renatus Murder Trial 1763-1764," item 15).
45. Sometimes spelled "Johann Arboe." Presumably his original family name was either "Arbø" or "Arbö."
46. Unless otherwise indicated, all biographical details about John Arbo are taken from his memoir (see Appendix).
47. Arbo referred to himself as "lame" and thus exempt from both county and provincial head taxes; see Arbo, J., ["John Arbo's Commonplace Book 1761-1765"], Feb., Nov. 1763. Lameness is not usually associated with severe cases of smallpox, and Arbo's portrait (fig. 12) does not show any facial scars from the disease. Perhaps Arbo had some other crippling disease such as polio, which was erroneously identified as smallpox. Alternatively, "lame" may only have been a convenient designation for the tax collector.
48. [Haidt, V.], "Johann Arboe," oil painting, [1760's] (MAB).
49. "Warden" was his official title in English; see "Ledger of the Diaconat at Bethlehem . . . ," E, ff. 89,166. His main function was as business manager for the Single Brethren's Choir.
50. Yates, W. R., et al., "Bethlehem of Pennsylvania. The First One Hundred Years," Bethlehem, 1968, pp. 101-104. Levering, pp. 376-385.
51. Arbo's books are easily identified by his signature on the title page. A listing of Arbo's library is available: "*Bücher von Johann Arbo ins Gemein-Archiv* [Books of John Arbo in the Congregation Archive]," no date, 2 pp. (MAB, file box "Bethlehem Archives and Library," folder 3). This list is attributed to John Ettwein (handwriting comparison) and may have been made at the time of Arbo's death in 1772.
52. Arbo, J., "*Des I. Theils der Mathematischen Anfangs-Gründe von der Arithmetic. Sectio I, von denen Numeris Absolutis.* [Part I, Mathematical Elements of Arithmetic. Section I, Absolute Numbers]," no date, 190 pp. (MAB, Arbo Collection, file box "Ms. Books of Joh. Arbo on Church History & Mathematics").
53. ["World Atlas"], bound collection of 80 folio maps, c. 1730-1760 (MAB, Arbo Collection).
54. "American Society. Minutes, 1766-68," Oct. 14, 1768, p. 131 (American Philosophical Society Library, Philadelphia, manuscript collection). Obituary in "*Wöchentliche Pennsylvanische Staatsbote*," Dec. 22, 1772, p. [2] (Library Company of Philadelphia).
55. "The American Philosophical Society . . . " in Pennsylvania "Gazette," Dec. 19, 1771, p. [1].
56. ["John Arbo's Commonplace Book"], 115 pp.
57. "*Conferenz Protocoll der Comittee in Bethlehem von 1762 bis d. 3te May 1780* [Conference Minutes for the Committee in Bethlehem from 1762 to May 3, 1780]," Nov. 5, 1764, p. 81 (MAB, file box "Bethlehem: *Conferenzen*. Journal of the Commission: 1752-1760. *Protocol der Comittee*: 1762-1780").
58. "*Pro Memoria der Bau der Oelmühle betr.* [Memorandum re Oil Mill Construction]," ["John Arbo's Commonplace Book"], Nov. 20, Dec. 12, 1764.
59. Unless otherwise indicated, the details of building the 1765 oil mill are taken from the "Oil Mill Diary" (see pp. 76-88).
60. "Oil Mill Diary," Jul. 8, Nov. 26, 1765. "Oil Mill Account Book," Jan. 24, Sep. 16, 1765. ["John Arbo's Commonplace Book"], Jan. 21, 1765. [Arbo, J., "Cost of Material and Work in Various Building Operations"], 1761-1770, p. [23] (MAB, file box "Bethlehem: *Bau-Inspection*. Building Material: 1761-1766 . . . "). "Ledger of the Diaconat at Bethlehem . . . ," E, f. 91, Jan. 21, 1765.
61. "Bethlehem Diary," Oct. 9, 1765, pp. 192-193.
62. Anonymous, "*Oel. Mühle in Bethlehem* [Oil Mill in Bethlehem]," drawing, [1766](MAB, Herrnhut Pictures 42.4; ABU, drawing NB-X O-II-42). We assume this drawing was made soon after the 1765 oil mill building was completed.
63. Arbo, J., "Notice is hereby given to the PUBLIC" in Pennsylvania "Gazette," Mar. 20, 1766, p. [4].
64. "Oil Mill Account Book," Oct. 10, 1767-Feb. 15, 1768.

65. Apparently little if any leather fulling was actually done at the 1765 oil mill. Only one reference to fulling leather there has been found in the extensive records available. A ledger entry for the first annual rental payment from the tannery reads "For bark stripping and leather stamping" (["Single Brethren's Ledger"], C, f. 124, May 31, 1767); but tannery payments in later years do not mention leather stamping. Annual financial reports for the Bethlehem cloth fulling mill (MAB, file box "*Abschlüsse mit den Hand Thierungen in Bethl.* 1764-1776") indicate that leather was fulled there from 1765 through 1776. It is not clear why leather fulling machinery was planned for the 1765 oil mill when similar facilities were already available and in use at the Bethlehem fulling mill.
66. "Oil Mill Account Book," Feb. 15, 1768.
67. ["Single Brethren's Ledger"], C, D, E, F, G, H & I, "*Oel-Mühle*" and related accounts, 1766-1814. These profit and loss data combine all the industries operated in the oil mill building: oilseed crushing, groat mill, tanbark mill, hemp mill, and snuff mill. 1767-1808 data were reported in pounds; 1809-1814 data were reported in dollars. £1 = $2.67 in 1808.
68. Litchfield, C., "Annual Commodity Inventories at the Bethlehem Oil Mill 1791-1813," 1 p., 1977 (MAB, file box "Single Brn. *Gewerbe Inventaria* 1794-1814").
69. "Oil Mill Diary," Dec. 28, 1765; Sep. 13-27, 1766. Presumably millet was stamped in the same mortars used to stamp flaxseed. See also: Fries, A. L. (ed.), "Records of the Moravians in North Carolina. Vol. I. 1752-1771," Raleigh, 1922, p. 125, Mar. 4, 1755.
70. ["Bethlehem Single Brethren's Journal"], Mar. 14, 1796; Mar. 18, 1797; May 31, 1797, p. 271 (MAB).
71. "Oil Mill Diary," Feb. 6, 20, May 7, 1769. [Arbo, J., "Cost of Material and Work in Various Building Operations"], 1761-1770. p. [43] (MAB, file box "Bethlehem: *Bau-Inspection*. Building Material: 1761-1766 . . . "). ["Single Brethren's Ledger"], C, f. 259.
72. ["Single Brethren's Ledger"], F, f. 244; G, ff. 76-78. "Oil Mill Diary," Sep. 29-Oct. 11, 1766.
73. "*Inventarium der Oelmühle* [Oil Mill Inventory]," May 31, 1794; May 31, 1795; May 31, 1796 (MAB, file box "Single Brn. *Gewerbe Inventaria* 1794-1814").
74. Butterfield, L. H., M. Friedlaender, and M. Kline (eds.), "The Book of Abigail and John. Selected Letters of the Adams Family, 1762-1784," Cambridge MA, 1975, p. 167.
75. "Bethlehem Diary," Jul. 25, 1782. For English translation see "Pennsylvania Magazine of History and Biography," vol. 13, p. 89, 1889.
76. Schoepf, J. D., "Travels in the Confederation (1783-1784)," translated and edited by A. J. Morrison, vol. 1, Philadelphia, 1911, p. 139.
77. La Rouchefoucault-Liancourt, [F. A. F.], "Travels Through the United States of North America, The Country of the Iroquois, and Upper Canada in the Years 1795, 1796, and 1797," 2nd ed., vol. IV, London, 1800, p. 130.
78. Comparison of the entries in the "Single Brethren's Choir Diary" from the 1760's with those of the 1790's reveals a major change in attitude among the Single Brethren. Religious experiences were emphasized in the earlier entries, while personnel problems were mentioned more frequently in the later decade.
79. Levering, pp. 593-599.
80. Lüders, T. C., "*Inventarium der Mobilien zur Oelmühle gehörig welche d. 31 May 1814 an Gideon Helwig mit der Mühle übergeben worden* [Inventory of the Portable Equipment Belonging to the Oil Mill That Was Transferred to Gideon Helwig with the Mill, May 31, 1814]," 1814, 1 p. (MAB, file box "Single Brn. *Gewerbe Inventaria* 1794-1814," folder "1814").

LOWER GRIST MILL AND WATERWORKS 1814-1934

1. For biographical information on Gideon Helwig (1770-1822) see his memoir (MAB, file box "Memoirs, Box 8, 1821-1830," folder "1822," no. 2).
2. Lüders, T. C., "*Inventarium der Mobilien zur Oelmühle gehörig welche d. 31 May 1814 an Gideon Helwig mit der Mühle übergeben worden* [Inventory of the Portable Equipment Belonging to the Oil Mill Which Was Transferred to Gideon Helwig with the Mill, May 31, 1814]," 1 p. (MAB, file box "Single Brn. House. *Gewerbe Inventaria*: 1794-1814," folder "1814").
3. "*Protokol des Aufseher Collegii in Bethlehem* [Minutes of the Bethlehem Supervisors Conference]," Mar. 21, 1814 (MAB). Bolger, S., "Waterworks Originally Oil Mill," BGT, Mar. 22, 1963, p. 6. Since the groat mill and later the water pumps occupied the eastern half of the oil mill building, this residence must have been built in the western half (see figs. 23, 28 & 31).
4. [Helwig, G.], "*Contra Buch für Gideon Helwig* [Separate Ledger for Gideon Helwig]," 1814-1823, 150 pp. + inserts (MAB, file box "Oil Mill: 1814-1823").
5. Lüders, T. C., E. Gehbe, S. Krause, and C. Luckenbach, "*Der Oelmühl Dürre Betreffend* [Concerning the Oil Mill Drying Equipment]," 4 pp., (MAB, file box "Oil Mill: 1814-1823"). "*Protokol* . . . ," May 23, 27, 1814.
6. "*Protokol* . . . ," Mar. 21, 1814.
7. For biographical information on Samuel Schultz (1794-1857) see: Myers, E., "Samuel Schultz, 1794-1857," BGT, Jun. 16, 1928, p. 6.
8. Anonymous, "Samuel Schultz 1798-1854," photograph, c. 1845, in Wolle, C. A., and M. C. Luckenbach, "Familiar Faces in Bethlehem, Pa. 1800-1900," 1915, p. 18 (MAB).
9. "*Samuel Schulz Dr. an Gideon Helwig Estat übernomen nach einem Presment den 28 January 1823* [Samuel Schultz Owes Gideon Helwig's Estate, Transferred According to an Appraisal, January 28, 1823]," 1 p. (MAB, file box "Single Brn. House. *Gewerbe Inventaria*: 1794-1814," folder "1823"). "*Inventarium der Mobilien zur Oehlmühle Gehörig welche den 14 Febr. 1823 an Samuel Schulz mit der Mühle übergeben worden* [Inventory of the Portable Equipment Belonging to the Oil Mill Which Was Transferred to Samuel Schultz with the Mill, Feb. 14, 1823]," 1 p. (MAB, file box "Inventories. Various Trades 1817-1826," folder "1823").
10. "*Protokol* . . . ," Jul. 14, 1834.
11. For biographical information on Charles D. Bishop (1784-1853) see: "Bethlehem Diary," Jul. 20, 1853, p. 257 (MAB); Myers, E. L., "John David Bishop Buried Nov. 5, 1827," BGT, Nov. 5, 1927, p. 6. The Bishop family was associated with the Bethlehem oil mill and waterworks building for well over a century. John David Bishop Sr. (1704-1763) drew the plans for the 1752 oil mill. His son John David Bishop Jr. (1749-1827) worked with millwright Hans Christensen. Gilbert Bishop (1817-1907), son of Charles D. Bishop and grandson of J. D. Bishop Jr., ran the grist mill facilities at the oil mill from 1863 to 1874. Augustus Bishop (1846-1910), a great-grandson of J. D. Bishop Jr., was chairman of the committee supervising waterworks operations while he served as a Bethlehem councilman (1880-1886).
12. Levering, J. M., "A History of Bethlehem, Pennsylvania 1741-1892," Bethlehem, 1903, p. 666. This drawing was made from a contemporary pen and pencil sketch by Rufus Grider (MAB, file box "Rufus Grider Coll. I," item 7).
13. Bishop, A. S., "Our Early Industries," BDT, Sesqui-Centennial Industrial Edition, Jun. 1892, p. 5.
14. Bethlehem Water Co., "TO LET! The lower Gristmill

in Water steet (*sic*) . . . ," broadside, 1846 (MAB, file box "Bethlehem Water Co. 1837-1850," folder "Printed Poster 1846").

15. Levering, p. 634.
16. Biographical details on Francis (Frantz) Zoller have not been located. Bishop (note 13) mentions him as a former tenant miller at the oil mill. Schultze ("The Old Moravian Cemetery of Bethlehem, Pa. 1742-1897" in "Transactions of the Moravian Historical Society," vol. 5, pp. 190, 209, 211, 245, 1899) lists Zoller's wife and three children and calls W. A. Zoller the "son of Henry (*sic*) Zoller, the oil miller."
17. "We, C. D. Bishop the successor to T. H. Goepp above said . . . ," Bethlehem, 1848, 2 pp. (MAB, file box "Bethlehem Water Co. 1837-1850," folder "Agreements 1838-1850"). "Statement of books, papers and effects delivered by J. C. Brickenstein former Secretary and Treasurer of the Bethlehem Water Company to Chas. W. Rauch his successor in office. April 13 & 16, 1850," p. [3] (MAB, file box "Bethlehem Water Co. 1837-1850," folder "Treasurer 1837-1850").
18. Henry, M. S., "History of the Lehigh Valley," Easton, 1860, p. 234. This probably was the Lewis Snyder who rented the mill in 1874-1875.
19. For biographical information on Gilbert Bishop (1817-1907) see: "Rev. Gilbert Bishop," SBG, Sep. 27, 1907, p. 1; "In Memoriam" in "The Moravian," Oct. 2, 1907, p. 632; Myers, E. L., "John David Bishop Buried Nov. 5, 1827," BGT, Nov. 5, 1927, p. 6.
20. Anonymous, "Rev. Gilbert Bishop 1817-1907," photograph, c. 1870, in Wolle, C. A., and M. C. Luckenbach, "Familiar Faces in Bethlehem, Pa. 1800-1900," 1915, p. 19 (MAB).
21. "*Protokol* . . . ," Jan. 25, 1819.
22. "More Pumping and Less Grinding Wanted," BDT, Nov. 23(misdated 22), 1872, p. 1. "Proceedings of Bethlehem Water Board," BDT, Aug. 4, 1874, p. 3.
23. ["*Bürgercasse* Ledger 1823-1846"], f. 90 (MAB, file box "Civil Treasury (*Bürgerliche Casse*). 2 Journals: 1823-1846"). "*Protokol* . . . ," Jul. 14, 1834; Feb. 9, 1835.
24. "More Pumping and Less Grinding Wanted," BDT, Nov. 23(misdated 22), 1872, p. 1. "Is Our Water Supply Sufficient?," BDT, Oct. 28, 1873, p. 3.
25. "The Water Board," BDT, Jan. 14, 1874, p. 3.
26. "Mr. Gilbert Bishop . . . ," BDT, Mar. 26, 1874, p. 3. "Rev. Gilbert Bishop," SBG, Sep. 27, 1907, p. 1.
27. Biographical details on Lewis Snyder have not been located.
28. "Proceedings of the Bethlehem Town Council," BDT, Jan. 17, 1872, p. 1. "Meeting of Bethlehem Water Board," BDT, Jan. 18, 1873, p. 1.
29. "Proceedings of Bethlehem Water Board," BDT, Aug. 4, 1874, p. 3. "The Old Moravian Mill," BDT, Aug. 28, 1874, p. 3. "Bethlehem Oat Meal," BDT, Sep. 30, 1874, p. 3. "Statement of the Treasurer of the Water Board," BDT, Mar. 22, 1875, p. 3.
30. "Proceedings of Bethlehem Water Board," BDT, Mar. 24, 1875, p. 3; Apr. 6, 1875, p. 3; May 4, 1875, p. 3. "Adjourned Meeting of the Bethlehem Water Board," BDT, Jun. 15, 1875, p. 3. "Specifications of Expense of the Water Board . . . ," BDT, Mar. 23, 1876, p. 2. These articles report the conversion of space in the grist mill into a residence for the waterworks engineer. Since the grist mill machinery located in the eastern half of the building was left intact (BDT, May 4, 1875, p. 3) and the water pumps were also situated on the eastern side (fig. 23), this project apparently consisted of renovating the 1814 miller's residence in the western half of the building.
31. Huetter, K. Z., "The Bethlehem Waterworks," Bethlehem, 1976, pp. 2-4.
32. At the spring house shown in figure 3.
33. "Mr. Charles D. Bishop, To Rush & Muhlenberg Dr., Philadelphia, September 8th, 1832," invoice, 1 p. (MAB, file box "Civil Treasury (*Bürgerliche Casse*) Bills etc.: 1832/33," folder B, item 49). Rau, R., "Historical Sketch of the Bethlehem Water Works," Bethlehem, 1877, p. 9.
34. "Bethlehem Water Board Meeting," BDT, Dec. 8, 1874, p. 3. "Proceedings of the Bethlehem Water Board," BDT, Jan. 5, 1875, p. 3.
35. "*Protokol* . . . ," Nov. 7, 1831. ["*Bürgercasse* Ledger 1823-1846"], ff. 85, 89 (MAB, file box "Civil Treasury (*Bürgerliche Casse*). 2 Journals: 1823-1846"). "*Bürgercasse übernommen von der Gemein Diaconie Gerathschaften in der Oelmühle* [Implements in the Oil Mill Transferred from the General Diacony to the Civil Treasury]," Bethlehem, 1834, 1 p. (MAB, file box "Civil Treasury (*Bürgerliche Casse*) Miscellanea 1803-1846: Accounting, Waterworks, Oil Mill, Taxes, etc.").
36. ["*Bürgercasse* Ledger 1823-1846"], f. 89. Cope, H. C., "Our Municipal History," BDT, Sesqui-Centennial Ind. Edition, Jun. 1892, p. 3. Roepper, W. T., "Draft of the Bethlehem Water Company's Property on Water Street & along Monockisy (*sic*) Creek," 1846, 1 p. (HBI, framed document; "Annual Report for 1962. The First National Bank & Trust Company of Bethlehem," back cover).
37. "Water!, Water!," BDT, Jun. 20, 1867, p. 3. "Bethlehem Water Works," BDT, Aug. 21, 1869, p. 2.
38. "Bethlehem Waterworks," BDT, Apr. 19, 1867, p. 3. "Water for Bethlehem," BDT, Jun. 27, 1867, p. 3. Cope, p. 4.
39. "The Bethlehem Water Company," BDT, Apr. 5, 1871, p. 1.
40. "The Water Question," BDT, May 17, 1871, p. 1. The purchase was made under authority of "An Act Authorizing the Borough Council of the borough of Bethlehem, Pennsylvania to purchase the water works . . . " enacted by the Commonwealth of Pennsylvania in 1866 and amended in 1871. For the text of this act see BDT, Jul. 12, 1871, p. 1.
41. Alderfer, E. G., "Northampton Heritage. The Story of an American County," Easton, 1953, pp. 304-305.
42. "Water!, Water!," BDT, Jun. 20, 1867, p. 3. "No Water," BDT, Aug. 19, 1867, p. 3.
43. "It Rained . . . ," BDT, Aug. 21, 1867, p. 3. "Water," BDT, Jan. 6, 1868, p. 3. "No Water," BDT, Mar. 18, 1868, p. 3. "The Reliance steamer . . . ," BDT, Apr. 30, 1874, p. 3.
44. ["Minutes of the Sessions of the Board of Managers of the Bethlehem Water Co."], session 10, Aug. 22, 1845; session 11, Sep. 2, 1845; session 14, Feb. 13, 1846 (MAB, file box "Bethlehem Water Co. 1837-1850," folder "Agreements 1838-1850"). "Statement of the books, papers and effects delivered by J. C. Brickenstein former Secretary and Treasurer of the Bethlehem Water Company to Chas. W. Rauch his successor in office. April 13 & 16, 1850," p. [3] (MAB, file box "Bethlehem Water Co. 1837-1850," folder "Treasurer 1837-1850"). Rauch, R. S., "Diary of Ruben S. Rauch, Feb. 24, 1854-March 7, 1857," Nov. 7, 1856 (MAB, file box "Bethlehem: Diary of R. S. Rauch 1854-1857"). Birkenbine, H. P. M., "Report on the Water Supply of Bethlehem," 1874, p. [1]. (MAB, file box "Bethlehem Water Supply ca. 1870"). Hunter, L. C., "A History of Industrial Power in the United States 1780-1930. Vol. 1: Waterpower in the Century of the Steam Engine," Charlottesville VA, 1979, pp. 313-318 (description of Howd's patent turbine).
45. The 1856 waterpower system with the undershot waterwheel apparently remained in service until 1888 when it was removed; see "Local News Notes. The

work . . . ," BDT, Oct. 26, 1888, p. 1. A replacement waterwheel was required in 1879. See: "Proceedings of Bethlehem Town Council," BDT, Apr. 17, 1879, p. 1; "The Repairs at the Water Works," BDT, May 7, 1879, p. 1; "Running Again," BDT, Aug. 13, 1879, p. 1.

46. "Water for Bethlehem," BDT, Jun. 27, 1867, p. 3.
47. "Water," BDT, Aug. 20, 1867, p. 3. "Water for Bethlehem!," BDT, Sep. 16, 1867, p. 3.
48. "Water for Bethlehem," BDT, Nov. 11, 1867, p. 3. "The Water Bubble Burst," BDT, Dec. 16, 1867, p. 3.
49. "Water," BDT, Jan. 6, 1868, p. 3.
50. "Trial of the New Pumping Engines," BDT, Jun. 8, 1868, p. 3.
51. Bailey, O. H., "Birds-Eye View of the Bethlehems, Pa.," Cincinnati, 1873, 1 p. (MAB, framed map; HBI, framed map). Note that the artist forgot to draw in the raceways for the oil mill.
52. "Inspecting the Water Works," BDT, Apr. 8, 1874, p. 4. "The Water Works," BDT, Apr. 10, 1874, p. 4.
53. Redrawn from contemporary insurance maps: "Bethlehem and South Bethlehem, Pennsylvania" (title varies with edition), Sanborn Map & Publishing Co., New York, 1885, f. 6; 1892, f. 10; 1897, f. 14; 1912 (c. 1920 pasteover underneath later blank pasteover), f. 22 (MAB). Original maps also show the number of floors and the type of wall and roof construction.
54. Eggert, "Second Water Works," photograph, c. 1880 (MAB, Thaeler Collection of Pictures, vol. TH-I, p. 18).
55. "The Reliance steamer . . . ," BDT, Apr. 30, 1874, p. 3. "At the Water Works," BDT, Jun. 10, 1874, p. 3. "Affairs at the Water Works," BDT, Jun. 29, 1874, p. 3. "Special Meeting of Bethlehem Water Board," BDT, Jun. 30, 1874, p. 3.
56. "The Steam Pump," BDT, Jul. 3, 1874, p. 3.
57. Grider, R. A., "No. 1. The first Water Works of Bethlehem & Powder Mag. in the Revolution. No. 2. The Oil Mill rebuilt present Water Works taken from the North," polychrome watercolor drawing, 1874 (MAB, file box "Rufus Grider Coll. I," item 7).
58. "Public Opinion," BGT, Dec. 7, 1933, p. 11.
59. "Proceedings of Bethlehem Town Council," BDT, May 5, 1885, p. 3.
60. "Last evening . . . ," BDT, Jun. 22, 1888, p. 4. "Our Water Supply," BDT, Aug. 31, 1888, p. 4. The tailrace of the Luckenbach grist mill ran within 30 feet (9 m) of the spring. The intervening red clay soil could not adequately filter any polluted water that might seep toward the spring.
61. "Borough Affairs," BDT, Aug. 13, 1888, p. 1. "Bethlehem Town Council," BDT, Oct. 2, 1888, p. 4.
62. "Bethlehem Town Council. The Water Question," BDT, Jul. 17, 1888, p. 1. "Town Council," BDT, Aug. 7, 1888, p. 1.
63. "Our Water Works," BDT, Oct. 24, 1888, p. 1.
64. "Proceedings of Bethlehem Water Board," BDT, Mar. 24, 1875, p. 3; Apr. 6, 1875, p. 3; May 4, 1875, p. 3. "Adjourned Meeting of the Bethlehem Water Board," BDT, Jun. 15, 1875, p. 3. "High Water," BDT, Aug. 14, 1875, p. 3. "Specifications of Expense of the Water Board . . . ," BDT, Mar. 23, 1876, p. 2. "Local News. J. Hartman . . . ," BDT, Apr. 21, 1885, p. 1.
65. "Bethlehem Town Council," BDT, Dec. 27, 1888, p. 1. "Local News Notes. The new stack . . . ," BDT, Mar. 23, 1889, p. 1. This article specifies a 60 foot chimney, while the Sanborn insurance maps (fig. 23) estimated it to be 70 (1892) and 90 (1897) feet high. "Bethlehem Water Works," BDT, Jan. 25. 1890, p. 4.
66. Anonymous, "South Face of Oil Mill & Dam," photograph, c. 1890 (HBI P-1774). We date this photo c. 1890 since the 1889 iron smokestack is present and the photo was published in Leibert, A. H., "Bethlehem Pa., Photo-Gravures," New York, 1894, p. [47] (BPL).
67. "North Side Town Council. The Water Works," BDT, May 7, 1889, pp. 1,4. "Bethlehem's Water Works," BDT, Oct 26, 1889, p. 1.
68. Rau, p. 10.
69. "Memorabilia of 1911 at Year's Final Service," SBG, Jan. 1, 1912, pp. 1, 6.
70. "Bethlehem Water Works," SBG, Dec. 7, 1911, p. 1.
71. "Water Supply Permit for Borough of Bethlehem," SBG, Mar. 8, 1912, p. 1. "Water Committee, Bethlehem Borough Council," SBG, Nov. 20, 1912, p. 1.
72. "Bethlehem Water Works," SBG, Dec. 7, 1911, p. 1. "New Centrifugal Pump," SBG, Nov. 7, 1912, p. 1.
73. "Bethlehem Borough Water," SBG, Jan. 10, 1913, p. 1.
74. "Oppose Razing Old Building," BGT, Dec. 6, 1933, pp. 1, 13 (because the oil mill had served as Bethlehem's waterworks since 1832, many people confused it with the nearby 1762 waterworks building). "Public Opinion," BGT, Dec. 7, 1933, p. 11; Dec. 8, 1933, p. 18. "Preserve Historic Building," BGT, Dec. 9, 1933, p. 6.
75. Phillips, C., "From the Crash to the Blitz 1929-1939," New York, 1969, pp. 268-269.
76. "Local Administrator Pushes Details to Take Scores Off Relief List . . . ," BGT, Nov. 21, 1933, pp. 1, 16.
77. Pfeifle, R., et al., "Fifth Annual Message of Mayor Robert Pfeifle to the Members of Council and Citizens of Bethlehem, Pa.," Bethlehem, 1934, p. 63.
78. "Brief History of Water Works Built by Moravians on North Side in 1754-55," BGT, Dec. 9, 1933, pp. 1, 10.
79. Pfeifle, R., et al., "Sixth Annual Message of Mayor Robert Pfeifle to the Members of Council and Citizens of Bethlehem, Pa.," Bethlehem, 1935, 282 pp. The 1932 (p. 275) and the 1934 (p. 230) annual reports describe the demolition of two other old mills owned by the city and reuse of the stone. However, the demolition of the oil mill is not mentioned in any annual report from the 1930's.
80. The WPA (Works Progress Administration) was the largest, the most costly, and the most conspicuous of the various work relief programs of the 1930's. As such, it came to represent all such programs in the public mind, regardless of whether a particular project was administered by the WPA, the Public Works Administration (PWA), the Civil Works Administration (CWA), the Civilian Conservation Corps (CCC), or some other agency. See Phillips, pp. 124, 256ff. The Bethlehem oil mill was demolished under the CWA program (note 76).
81. Gill, B. C., "The Tawry (*sic*) and Springhouse, Project B-66," unpublished HBI report, Bethlehem, [1975], pp. 16-18 (HBI). Describes excavations in the oil mill headrace area. The only known records of the dig at the oil mill foundation are archeology slides CS-68-259 through CS-68-275 in the HBI collection and four slides in the personal collection of Stephen Young.
82. Gilman, T., "Site of Bethlehem Oil Mill and Restored Waterworks," photograph, 1980.

TECHNOLOGY OF THE OIL MILL INDUSTRIES - I

1. [Schulze], "*Grund Riß der Oel-Mühle in Bethlehem* [Plan Drawing of the Oil Mill in Bethlehem]," [1766], 1 p. (ABU, drawing NB-X O-II-43; MAB, Rinker films, roll 49). The original 1766 Herrnhut drawings contain some obvious errors: the oilseed stamper guides on the south side of the main shaft were omitted from the second floor plan (fig. 28); the distance between the base log of the tanbark mill and the large adjacent face gear is different in the plan (fig. 29) and elevation (fig. 31) drawings; etc.
2. "Oil Mill Account Book," Nov. 23, 1767 through Feb. 15, 1768. "Oil Mill Diary," Dec. 15, 1767 through Feb. 5, 1768.

3. [Schulze], "*Stand Riß der Oel-Mühle in Bethlehem* [Elevation Drawing of the Oil Mill in Bethlehem]," [1766], 1 p. (ABU, drawing NB-X O-II-44; MAB, Rinker films, roll 49). See comment in note 1.
4. Erbe, H., "*Bethlehem, Pa. Eine kommunistische Herrnhuter Kolonie des 18. Jahrhunderts*," Stuttgart, 1929, 191 pp. (MAB). Unpublished English translation: Erbe, H., "Bethlehem, Pa., A Communistic Herrnhut Colony of the 18th Century," [Bethlehem, 1959], p. 69, figs. VIa, VIb (HBI, MAB). All our page references refer to the English translation. Close comparison of Erbe's redrawing of the oil mill plans with the original 1766 Herrnhut drawings shows numerous discrepancies in mechanical details.
5. "Oil Mill Account Book," Aug. 31, 1766.
6. "Oil Mill Diary," Aug. 23, 24, Sep. 16, 1765.
7. "Available net head" is the difference in elevation from the top surface of the headwater to the top surface of the tailwater. Rau claimed a 2 foot head for the 1762 waterworks wheel ("Historical Sketch of the Bethlehem Water Works," Bethlehem, 1877, p. 5), and the head at the 1765 oil mill must have been nearly identical (figs. 56 & 57). In 1879 there was "22 inches of water fall" at the oil mill dam ("Proceedings of Bethlehem Town Council," BDT, Apr. 22, 1879, p. 1).
8. Daugherty, R. L., and J. B. Franzini, "Fluid Mechanics with Engineering Applications," 6th ed., New York, 1965, pp. 374-375. Church, I. P., "Hydraulic Motors with Related Subjects," New York, 1907, pp. 35-36. Calculation by Stephen Young making certain necessary assumptions concerning water flow and wheelpit geometry.
9. Jesperson, A., "*Svanninge Stampemølle.* Reconstructing the Fulling Mill at *Købstadsmuseet Den Gamle By*, Aarhus," Copenhagen, 1966, p. 65.
10. Leupold, J., and J. M Beyern, "*Theatrum machinarum molarium, oder Schau-Platz der Mühlen-Bau-Kunst* [A Theater of Mill Machinery]," 2nd ed., Dresden, 1767, part I, pp. 81-85, plates XXVIII, XXIX. Linperch, P., "*Architectura Mechanica, of Moolen-Boek* [Mechanical Architecture, or Mill Book]," Amsterdam, 1727 (reprinted Amsterdam, 1969), pp. 1-2, plates 1-6. Sipman, A., "*Molentekeningen* [Mill Plans]," Zutphen, Netherlands, 1978, plates 24-33. Kempers, A. J. B., "*Oliemolens* [Oil Mills]," Arnhem, 1962, 92 pp. Weber, F. W., "*Die Historischen Ölmühlen der Pfalz* [Historic Oil Mills of the Pfalz]" in "*Mitteilungen des historischen Vereins der Pfalz*," vol. 75, pp. 191-225, 1977.
11. "Extract from the Minutes of the American Society, held at Philadelphia, for promoting useful Knowledge. The Committee appointed to examine the letters and accounts, relating to the expression of oil from the sunflower seeds, report as follows" in Pennsylvania "Gazette," Oct. 6, 1768, p. [2].
12. Breitenbach, F., "*Die Oel-Oekonomie* [The Oil Economy]," Berlin, 1806, pp. 98-99.
13. Figure 14. [Arbo, J., "Cost of Material and Work in Various Building Operations"], 1761-1770, p. [42] (MAB, file box "Bethlehem: *Bau-Inspection.* Building Material: 1761-1766 . . . "). English translation: "In the year 1767 we did not clean the flaxseed beforehand but only did the coarse sieving, and the oilseed cakes were made thinner and remained in the press longer. Then we obtained a good quart of oil per bushel more than before, i.e. 7 quarts of oil per bushel [of seed]." This indicates the normal use of two procedures for cleaning flaxseed at the oil mill: one to remove fines and the other to remove coarse impurities. ["Single Brethren's Ledger"], C, f. 259 (MAB). English translation: "There was also a bin constructed which was not included in the account. A wind mill [fanning machine] and two brass sieves for cleaning split peas and flaxseed in the clapper mill were also left out."
14. "Bethlehem Diary," Oct. 9, 1765, pp. 192-193 (MAB). Truman, R., "MADE and SOLD by RICHARD TRUMAN . . . Wheat fans for cleaning all sorts of grain and flaxseed" in Pennsylvania "Gazette," Oct. 6, 1768, p. [8].
15. More mortars would have been required for stamping loose flaxseed than for the compacted press cake, so the southern mortars were probably used mostly for the seed. The mortars on the northern side are conveniently adjacent to the mill for grinding the cake, so we assume they were mainly used for stamping the ground press cake.
16. Details not given in the 1766 Herrnhut drawings are taken from similar European oilseed mills of the period (note 10). For clarity, we depict the stampers on the northern side of the oilseed mill because they have the camshaft and side arms located on the second floor (fig. 28). The southern stampers, the ones most likely used for seed crushing (note 15), were lifted by identical cams located on the waterwheel shaft on the first floor (fig. 29).
17. "Oil Mill Diary," Jan. 18, 1766.
18. "Oil Mill Account Book," p. [64], section "Summary of Costs. Miscellaneous."
19. Details not given in the 1766 Herrnhut drawings are taken from similar European oilseed mills of the period (note 10).
20. A "pan" was embedded in masonry at the oil mill, presumably the pan for roasting crushed flaxseed. "Oil Mill Diary," Feb. 3, 1766. "Oil Mill Account Book," Feb. 22, 1766.
21. The position of this flue corresponds with the chimney shown in fig. 14.
22. No connection between the stirrer shafts and the waterwheel power train is shown in the 1766 drawings, but a rope drive is implied. Near the eastern end of the overhead horizontal shaft in the stirring mechanism (fig. 29) is a wheel or pulley that lines up with the rope drive pulley on the upper cam shaft for the oilseed stampers (figs. 30 & 31). This camshaft pulley, which also powers the clapper mill with a rope drive, appears to have a double groove in fig. 28.
23. "Oil Mill Account Book," Feb. 22, 1766; p. [64], section "Summary of Costs. Miscellaneous."
24. "Oil Mill Account Book," Feb. 1, 1766; p. [64], section "Summary of Costs. Miscellaneous," line "To Just for . . . "
25. "Oil Mill Diary," Mar. 16, 1768.
26. "Oil Mill Diary," Apr. 3, 1766; Feb. 28, Nov. 9, 1767; May 5, 1768; Jun. 1, Oct. 26, 1769.
27. "Oil Mill Diary," Apr. 3, 1766. [Arbo, J., "Cost of Material . . . "], p. [42].
28. "Oil Mill Diary," Jun. 5, 1769.
29. Otto, [J. M.], "The Letter from Doctor Otto to Doctor Bond, gives the following account of an Oil, made from the seeds of the common large Sunflower" in Pennsylvania "Gazette," Oct. 6, 1768, p. [2]. Although not specifically stated, Otto must have carried out his experiments at the Bethlehem oil mill. His apothecary shop was located only 350 feet (107 m) from the mill, and he states that "The [sunflower] oil is made much in the same manner as the linseed oil." Although sunflowers were grown in colonial Pennsylvania (note 11), it was too far north for cotton. Perhaps Otto obtained his cottonseed from the Moravian settlement at Salem, North Carolina.
30. Erbe, p. 71.
31. Gehret, E. J., and A. G. Keyser, "Flax Processing in Pennsylvania from Seed to Fiber" in "Pennsylvania Folklife," vol. 22, no. 1, pp. 10-34, 1972.

32. The Moravians at Bethlehem used dew retting rather than water retting for their flax, so we assume that they also used dew retting for their hemp as was common in colonial Pennsylvania. See Jordan, J. W., "Bethlehem During the Revolution" in "Pennsylvania Magazine of History and Biography" vol. 13, p. 73, 1889.
33. Weber, F. W., personal communication. The early Pennsylvania flax kilns described by Gehret & Keyser (note 31) probably also served for drying retted hemp.
34. "Hemp, and the Early Hemp Mills of Lancaster County" in Lancaster "Sunday News," Sep. 2, 1928, p. 14. Wright, J., "Hemp Growing Long Ago" in Lancaster "New Era" (semi-weekly edition), Jun. 24, 1905, p. 6. Hall, R., "Observations Made by Richard Hall, of the City of Dublin, Hemp and Flax Dresser; on the Methods Used in Holland, in Cultivating or Raising of Hemp and Flax," Dublin, 1724, pp. 9-37. Wissett, R., "A Treatise on Hemp," London, 1808, pp. 226, 230, 268-269. Butowsky, A., "The Means of Improving the Management, Cultivation, and Fabrication of Hemp," St. Petersburg, Russia, 1842; English translation by P. von Schmidt in "The New American State Papers. Agriculture," Wilmington, 1973, vol. 2, pp. 436-499. Weber, F. W., "*Die Geschichte der pfälzischen Mühlen besonderer Art* [The History of Special Types of Mills in the Pfalz]," Otterbach, West Germany, 1981, pp. 215-229 (MAB).
35. An undated sketch of water-powered hemp/flax braking and scutching machinery exists at the Moravian Archives (MAB, map & print file, drawer II, folder 5, no. 10, copy 20A). We have seen no evidence that such machinery was ever built in Bethlehem.
36. The camshaft for the hemp stamping mill is shown floating in midair in the 1766 Herrnhut drawings (fig. 30). Apparently the draftsman deliberately omitted the bearings and drive gears for this shaft, possibly for clarity in depicting the other machinery.
37. Leupold & Beyern, pp. 102-103, plate XXXIV (fulling mill). Contemporary European hemp and flax stamping mills commonly used flat-bottomed stampers. See: Schilli, H., "*Der Vogtsbauernhof* [The Government Administrator's Farmstead]," Lahr/Schwarzwald, 1975, pp. 63-64, 88; van Natrus, L., C. van Vuuren, and J. Polly, "*Groot Volkomen Moolenboek* [The Great Complete Mill Book]," Amsterdam, 1736 (reprinted Amsterdam, 1969), plate X; Jespersen, A. (ed.), "Report on Watermills. Vol. 3. Scale Drawings," Virum, Denmark, 1957, mill D-50001, plate "Section of flax stamp, from south." The unconventional use of stepped heads at Bethlehem may have been dictated by the intended dual use of the machinery for both hemp stamping and leather fulling. A contemporary sketch of a hemp stamper using flat-bottomed stampers has been found at the Moravian Archives, but there is no evidence that such machinery was ever built (MAB, map & print file, drawer II, folder 5, no. 11, copy 29B; Weber, p. 221).
38. "House Conference 1750-[1752]," Mar. 29, 1752 (MAB, file box "Single Brn., House Conference: 1750-1752").
39. Unlike the wet process for fulling wool, the hemp fiber was kept dry during stamping (Butowsky, pp. 463-464).
40. "Bethlehem Diary," Oct. 9, 1765, pp. 192-193. Levering, J. M., "A History of Bethlehem, Pennsylvania 1741-1892," Bethlehem, 1903, p. 410.
41. "Oil Mill Account Book," p. [47], Nov. 23, 1767 through Feb. 15, 1768. Note the expenditures for a roller stone, an iron axle for that stone, a base or platform, and an iron collar around that base.
42. "Hemp, and the Early Hemp Mills of Lancaster County" in Lancaster "Sunday News," Sep. 2, 1928, p. 14. Old mill stones shaped like a truncated cone and having a central axle hole are found in several Pennsylvania museum collections: Mercer Museum (Doylestown) (fig. 63), Pennsylvania Farm Museum (Lancaster), Colonial Valley (Menges Mills), Hans Herr House (Willow Street).
43. Schilli, pp. 88-89.
44. Litchfield, C., "Hemp Roller Mill, *Vogtsbauernhof*, Gutach, Germany," photograph, 1979 (HBI P-7125).
45. Butowsky, p. 463.
46. The German name "*Hanfreibe*" (literally "hemp rubber") was applied to both the hemp stamping mill and to the hemp roller mill. Both performed the same hemp-softening function, even though they were entirely different types of machinery.
47. The presence of two different types of hemp mills in one building also raises the possibility that one might have been used for softening hemp and the other for a second purpose such as braking hemp. We believe that both mills were used for softening hemp since both mills have the same German name "*Hanfreibe*" and Arbo's 1766 ad (fig. 15) directly compares the results from the two types of mills.
48. ["Single Brethren's Ledger"], C, f. 78; D, f. 49. There is no indication of how production was divided between the hemp stamping and hemp roller mills during 1770-1772.
49. ["Single Brethren's Ledger"], G, f. 74, Jan. 18, 1792.
50. Erbe, pp. 65-66 (translation revised slightly).
51. After the 1752 oil mill burned in 1763, a temporary bark mill (location unknown) was constructed in the spring of 1764. Its technology is unknown, but a crushing roller stone would have been the cheapest alternative. Presumably this temporary tanbark mill was abandoned in 1766 when the new tanbark stamping mill at the 1765 oil mill building was started up. See "*Haupt-Buch vor die Bau Inspection in Bethlehem 1762-1766* [Ledger for Building Supervision in Bethlehem 1762-1766]," f. 62 (MAB, file box "Bethlehem: *Bau Inspection. Haupt-Buch*: 1762-66 & 1766-1803").
52. Ellsworth, L. F., "The American Leather Industry," Chicago, 1969, pp. 5-7.
53. Weiss, H. B., and G. M. Weiss, "Early Tanning and Currying in New Jersey," Trenton, 1959, pp. 16-22, 32-34. Sprague, W. B., "The Tanner and the Currier" in "Chronicle of the Early American Industries Association," vol. 1, pp. 14-4, 14-6, 15-3, 15-4, 16-3, 1935. Ellsworth, pp. 5-7. Welsh, P. C., "Tanning in the United States to 1850," Washington, 1964, pp. 15-35.
54. de Fontenelle, J., and F. Malepeyre, "The Arts of Tanning, Currying, and Leather-Dressing," translated by C. Morfit, Philadelphia, 1852, p. 210.
55. Weiss & Weiss, pp. 26-27. Elsworth, pp. 7-9. "Oil Mill Diary," Sep. 1, 1766.
56. "*Haupt-Buch vor die Roth-Gerberey in Bethlehem 1765* [Ledger for the Bethlehem Tannery 1765]," pp. 25, 27, 29, 31 (MAB, file box "*Roth-Gerberei*"). An inventory of 100 cords (363 cubic meters) was recorded in May 1781 and May 1782. See "*Rothgerberey Inventarium* [Tannery Inventory]," 1781, p. 1; 1782, p. 1 (MAB, file box "[General Diacony], Inventories, Various Trades 1781-1785").
57. Landis, H. K., "Tan-Bark Mills" in "Chronicle of the Early American Industries Association," vol. 1, p. 24-8, 1937.
58. Welsh, p. 8. Weiss & Weiss, pp. 29-30. Landis, H. K., pp. 24-7, 24-8.
59. Leupold & Beyern, plate XXXI. Moravian millwrights apparently preferred this type of bark mill. Drawings exist for a wind-powered tanbark stamping mill at

the Moravian village of Kleinwelke, Germany (near Herrnhut). See "*Entwurff zu einer Loh-Mühle vom Winde getrieben mit einem bewechlichen Dache* [Design for a Wind-Driven Tanbark Mill with a Movable Cap (Roof)]," 1788, 2 pp. (MAB, Rinker films, roll 51; ABU, drawings IV-22a, IV-22b).

60. "Oil Mill Account Book," p. [66], section "Summary of Costs. Ironwork." Purchased "4 irons with rings for the bark stamper." Apparently the stamping heads ("irons") were attached to the ends of the poles using shrink bands ("rings"). Leupold & Beyern, plate XXXI.

61. "Oil Mill Account Book," Aug. 31, 1766. The wooden case around the base of the tanbark stampers probably looked like the "wooden stamping box" for the snuff mill shown in fig. 48 or like the cabinet built around the stampers of the spice mill illustrated in Leupold & Beyern, plate XXXI.

62. "Oil Mill Diary," Jun. 28, Aug. 19, Sep. 11-13, Sep. 29-Oct. 11, 1766. ["Single Brethren's Ledger"], C, f. 124. Since the volume of each mortar was c. 3.5 bushels (figs. 29 & 31) and the ledger counts 3 bushels per mortarload, we assume the ledgerkeeper was measuring raw bark rather than the denser tanbark powder. Annual production capacity = (20 mortarloads per day)(300 days).

63. "*Inventarium der Oelmühle* [Oil Mill Inventory]," May 31, 1794 (MAB, file box "Single Brn. *Gewerbe Inventaria* 1794-1814").

64. "*Protokoll Buch des Aufseher Collegii in Bethlehem* [Minutes Book of the Bethlehem Supervisors Conference], Jan. 13, Feb. 3, 17, Nov. 3, 17, 24, Dec. 8, 1806; Feb. 16, 23, Aug. 3, 10, Sep. 14, 21, Oct. 12, 1807 (MAB).

65. ["General Diacony Journal"], C, p. 171, Apr. 26, 1808 (MAB). Note also that annual payments from the tannery to the oil mill for use of the tanbark mill ended in 1807 (["Single Brethren's Ledger"], I, f. 65, May 31, 1805, Apr. 30, May 31, 1806; f. 126, May 30, 1807 (MAB).

TECHNOLOGY OF THE OIL MILL INDUSTRIES - II

1. The "oat meal" of eighteenth century Bethlehem was a coarsely ground meal. Modern American "oatmeal" is not a ground product at all but is "rolled oats" produced by flattening hulled oat kernels between steel rollers.

2. Fig. 14. "Oil Mill Diary," Dec. 28, 1765; Sep. 2, 3, 13-27, 1766. "Bethlehem Diary," Oct. 9, 1765, pp. 192-193 (MAB). Stamped and presumably winnowed millet was used directly for porridge by the Moravians in North Carolina; see Fries, A. L., (ed.), "Records of the Moravians in North Carolina. Vol. I. 1752-1771," Raleigh, 1968, p. 125, Mar. 4, 1755.

3. Weber, F. W., "*Die Geschichte der Mühlen und des Müllerhandwerks der Pfalz* [The History of Mills and the Milling Trade in the Pfalz]," Otterbach, West Germany, 1978, pp. 33-35. Leupold, J., and J. M. Beyern, "*Theatrum machinarum molarium, oder Schau-Platz der Mühlen-Bau-Kunst* [A Theater of Mill Machinery]," 2nd ed., Dresden, 1767, part I, pp. 85-86 (EMHL).

4. "Oil Mill Diary," Mar. 3, 1766.

5. The "Oil Mill Diary" makes no mention of stamping barley after the groat mill stones started operating late in 1766. Millet stamping is mentioned in Arbo's 1766 advertisement (fig. 15) and in the "Bethlehem Diary" (Oct. 9, 1765, pp. 192-193) as though it were a permanent process at the oil mill.

6. "Oil Mill Diary," Sep. 29-Oct. 11, 21, Nov. 13, 24-26, Dec. 10, 20, 22, 1766.

7. Murtagh, W. J., "Moravian Architecture and Town Planning," Chapel Hill NC, 1967, pp. 32-34.

8. Litchfield, C., "Annual Commodity Inventories at the Bethlehem Oil Mill 1791-1813," 1 p. (MAB, file box "Single Brn. House. *Gewerbe Inventaria* 1794-1814").

9. "Oil Mill Account Book," Jan. 29, 1767. "Bethlehem Diary," Oct. 9, 1765, pp. 192-193. Lüders, T. C., "*Inventarium der Mobilien zur Oelmühle gehörig welche d. 31 May 1814 an Gideon Helwig mit der Mühle übergeben worden* [Inventory of the portable equipment belonging to the oil mill which was transferred to Gideon Helwig with the mill, May 31, 1814]," 1814, 1 p. (MAB, file box "Single Brn. House. *Gewerbe Inventaria* 1794-1814," folder "1814").

10. Redrawn from Storck, J., and W. D. Teague, "Flour for Man's Bread," Minneapolis, 1952, p. 190. Some early German fanning machines were built without screens and performed only the winnowing function (*Vogtsbauernhof Freilichtmuseum*, Gutach, West Germany). The fanner at the Bethlehem groat mill probably did contain sieving screens, since considerable brass wire was purchased for its construction ("Oil Mill Account Book," Jan. 29, 1767).

11. ["Single Brethren's Ledger"], C, f. 259 (MAB).

12. *Klappermachine* was a common German name for a winnowing machine that operated with a repetitive clapping noise (Eger, G., and O. Brandes, "Technological Dictionary in the English and German Languages," vol. 2, Braunschweig, 1884, p. 449), but we have not found a design similar to the Bethlehem clapper mill in the German mill literature. Arbo speaks about grinding *Graupen* [peeled barley] to produce three products: *Perl Graupen* [pearl barley], *schlechtere* [poorer quality], and *Mehl* [meal]; see [Arbo, J., "Cost of Material and Work in Various Building Operations"], 1761-1770, p. [42] (MAB, file box "Bethlehem: *Bau-Inspection*. Building Material: 1761-1766 . . . "). Therefore, we assume that the clapper mill separated three fractions: coarse, medium, and fine. The "Oil Mill Account Book" records the purchase of sheet metal sieves (Mar. 31, Oct. 21, Nov. 29, 1766) and a wooden box (May 14, 1766; p. [66]) for the clapper mill. The cam/pivot/trip-hammer mechanism is interpreted from the top view of the clapper mill drawn in 1766 (fig. 28). Oscillating sieves were commonly found in contemporary European groat mills: Linperch, P., "*Architectura Mechanica, of Moolen-Boek* [Mechanical Architecture or Mill Book]," Amsterdam, 1727 (reprinted Amsterdam, 1969), figs. 18, 19; Leupold & Beyern, plates XXX, XXXI; Kempers, A. J. B., "*In en om de Grutterij* [In and Around the Groat Mill]," Arnhem, 1979, pp. 37-39.

13. Arbo (p. [53]) speaks of taking oats directly from the kiln to the hulling mill, so apparently the drying step came after cleaning. Craik, D., "The Practical American Millwright and Miller," Philadelphia, 1877, pp. 353-356. Jones, D. H., "The Water-Powered Cornmills of England, Wales, and the Isle of Man. A Preliminary Account of Their Development" in "Transactions of the International Molinological Society," vol. 2, Lyngby, Denmark, 1969, pp. 331-333. Jones, D. H., and J. K. Major, "Drying Kilns Attached to British Corn Mills," ibid., vol. 3, Netherlands, 1973, pp. 135-141.

14. The oat drying kiln was "18 feet [5.5 m] long, 12 feet [3.7 m] wide, 5 feet [1.5 m] in the ground, 10 feet [3.0 m] above the ground." It was built of stone and brick and had a gabled roof. A wooden shed was attached on the north side. See: fig. 3; "Oil Mill Account Book," Oct. 10 through Dec. 12, 1767; Arbo, p. [41]. According to the 1814 plans for redesigning the heating apparatus, the fireplace was located at the center of the funnel-shaped underground portion of the kiln; see Lüders, T. C., E. Gehbe, S. Krause, and C. Luckenbach, "*Das Oelmühl Dürre betreffend*

[Concerning the Oil Mill Drying Equipment]," 1814, 4 pp. (MAB, file box "Oil Mill: 1814-1823").

15. "*Chor-Diarium der Ledigen Brüder* [Single Brethren's Choir Diary]," Aug. 14, 1773 (MAB).
16. Arbo (p. [45]) speaks of converting unhulled barley into peeled barley by running it through the hulling stones. Ledger entries for the groat mill mention "rolling barley" [*Gerste gerollt*], presumably to produce peeled barley; see ["Single Brethren's Ledger"], C, f. 58, Jan. 17, Feb. 14, 28, 1767. Weber (p. 78) also mentions the German use of hulling stones to "roll" barley.
17. "Oil Mill Account Book," Aug. 31, 1766. "Oil Mill Diary," Sep. 29-Oct. 11, 21, Nov. 13, 24-26, Dec. 10, 20, 22, 1766.
18. This calculation is based on a waterwheel speed of 7.5 rpm, a value derived from the power calculations (p. 43). Gear ratios were determined by measuring the pitch diameters of the gears shown in the 1766 Herrnhut drawings. These diameters are presumed to be accurate, although it appears that the number of teeth on each gear was reduced for ease of drawing.
19. Redrawn from the 1766 Herrnhut drawings (figs. 30 & 31) adding additional details from typical Germanic millstone installations of the period: Krünitz, J. G., "*Ökonomisch-technologische Encyklopädie* [Economic and Technological Encyclopedia]," vol. 95, Berlin, 1804, pp. 360-365, fig. 5642; Schilli, H., "*Die Schwarzwälder Mühle* [The Black Forest Mill]" in "*Ekkhart-Jahrbuch 1966 der Badischen Heimat* [Ekkhart 1966 Yearbook of the Baden Homeland]," Freiburg, 1966, pp. 1-28; Wittich, R. (ed.), "*Romantik und Wirklichkeit der alten Mühlen* [Romance and Reality of the Old Mills]," 2nd ed., Kassel, 1980, pp. 124-125; and citations below. The distinctive one-stone *Perl-Graupen-Muhle* [pearl barley mill] design so frequently mentioned in the German mill literature (Leupold & Beyern, pp. 85-86, plates XXX, XXXI) was not employed at the Bethlehem groat mill, since the "Oil Mill Account Book" (Aug. 27, 1765) records the purchase of four millstones for the groat mill. It was not uncommon for the bedstone to be much larger than the runner stone and for the case to cover only the runner stone in German grist mills of the eighteenth century (Schilli, H., "*Der Vogtsbauernhof* [The Government Administrator's Farmstead]," Lahr/Schwarzwald, 1975, pp. 62-64; Jordan, A. F., "Some Early Moravian Builders in America" in "Pennsylvania Folklife," vol. 24, no. 1, Lancaster, 1974, p. 11). However, the 1766 Herrnhut drawings indicate that this design was not followed in the Bethlehem groat mill.
20. Weber, p. 73. Craik, pp. 356-359.
21. The "Oil Mill Diary" (Oct. 23-28, 1769) specifically refers to Christensen working on the hulling mill (*Schäl Mühle*). Arbo (p. [53]) speaks of putting oats first through the spelt mill (presumably the hulling mill) and then through the grinding mill.
22. The conversion of peeled barley to pearl barley consisted of grinding down the outer surface of the barley kernel without cracking the kernel. The hulling stones would seem better suited for such an attrition process than the grinding stones. However, we have not found any information indicating which stones the Moravians used to produce pearl barley.
23. The 1766 Herrnhut drawings (figs. 28-31) show no indications of a hopper to feed the clapper mill. Christensen installed "grist chutes" at the the groat mill ("Oil Mill Diary," Oct. 21, 1766) after these drawings had been completed. Perhaps these chutes conveyed grist from the grinding stones to the clapper mill.
24. Arbo, pp. [42-45]. All yields have been converted to a common basis: bushels of product per ten bushels of original unhulled grain.
25. We have not found data that would permit us to estimate production levels at the groat mill. Although a separate groat mill account was maintained in the Single Brethren's Ledger from 1766 to 1777, the product sales recorded seem unreasonably low compared to the groat consumption one would expect for a community the size of Bethlehem. Many large sales are to Nicholas Garrison for transport to the West Indies. Apparently this groat mill account lists mainly "external" transactions. See ["Single Brethren's Ledger"], C, ff. 58-60, 297; D, ff. 41-42, 204, 248; E, ff. 59, 137-138.
26. Fig. 28, caption "*Speltz und Graupen Mühlen* [Spelt and Groat Mills]." "*Inventarium der Oelmühle* [Oil Mill Inventory]," May 31, 1792, May 31, 1804, May 31, 1812, May 31, 1813 (MAB, file box "Single Brn. House. *Gewerbe Inventaria* 1794-1814"). "Oil Mill Diary," Jan. 26, 1767. "Oil Mill Account Book," Oct. 21, 1766. Fleischmann, C. L., "Result of observations made during a visit to Germany in 1844-45" in "Annual Report of the Commissioner of Patents for the Year 1847," Washington, 1848, pp. 336-338; reprinted in "The New American State Papers. Agriculture," Wilmington, 1973, vol. 5, pp. 352-354. Since husked spelt was customarily ground into flour like normal wheat in Germany, we assume that the Moravians husked their spelt at the groat mill but then took it to their nearby flour mill for grinding.
27. "Oil Mill Diary," Oct. 21, 1766. "*Inventarium der Oelmühle*," May 31, 1795, May 31, 1803, May 31, 1804.
28. "Single Brethren's Choir Diary," Mar. 17, 1768. "Oil Mill Diary," Feb. 6, 20, May 7, Jun. 12, Nov. 13, 1769; Jan. 2, Feb. 5, Oct. 30, 1770. The exact process used to make split peas at the groat mill is unknown. It may have been similar to that described by Craik (pp. 364-366) a century later: (a) soak the shelled peas in cold water until they swell and burst the hull, (b) dry them in a hot grain kiln without smoke, (c) split and hull the peas with the same hulling stones used for cereal grains, and (d) blow away the hulls with a fanning machine.
29. ["Bethlehem Single Brethren's Journal"], 1793-1798, ff. 194, 255, 271 (MAB).
30. "*Inventarium der Oelmühle*," May 31, 1794. This "*Habermehl Beutel & Kasten* [oat meal sieve (or bolter) and box]" may have been just a simple sieving device rather than a bolter. It cost only £ 4.18.-, and an 1814 inventory of oil mill equipment lists a "*Kasten u. Habermehlsieben* [box and oat meal sieves]"(note 9).
31. "*Protokol des Aufseher Collegii in Bethlehem* [Minutes of the Bethlehem Overseers Conference]," May 23, June 27, 1814. Lüders, T. C., E. Gehbe, S. Krause, and C. Luckenbach, "*Das Oelmühl Dürre betreffend* [Concerning the Oil Mill Drying Equipment]," 1814, 4 pp. (MAB, file box "Oil Mill: 1814-1823").
32. Bishop, A. S., "Our Early Industries," BDT, Sesqui-Centennial Industrial Edition, Jun. 1892, p. 5. "Bethlehem Oat Meal," BDT, Sep. 30, 1874, p. 3.
33. "Proceedings of Bethlehem Water Board," BDT, Apr. 6, 1875, p. 3.
34. An 1875 inventory of milling equipment still at the oil mill/waterworks mentions a "patent mill" ("What the Borough Owns," BDT, Nov. 24, 1875, p. 3). Nothing is known about how it was used.
35. "*Inventarium der Tobacks Fabrique* [Inventory of the Tobacco Manufactory]," 1791-1813, (MAB, file box "Single Brn., *Gewerbe Inventaria* 1794-1814"). ["Single Brethren's Ledger"], F, f. 244; G, ff. 76-78, 206-207; H, 108-109; I, ff. 63, 139, 175.
36. "Oil Mill Diary," Sep. 29-Oct. 11, 1766. [Helwig, G.],

"*Contra Buch für Gideon Helwig* [Separate Ledger for Gideon Helwig]," numerous ledger entries for grinding tobacco between 1814 and 1823 (MAB, file box "Oil Mill: 1814-1823").

37. Since the actual formulas for making snuff at the Bethlehem tobacco manufactory have not survived, this flow chart is based upon typical German, English, and American snuff manufacturing practices of the period. There were, of course, many proprietary variations of the seven basic processing steps shown here. For example, some snuff makers milled their tobacco before rather than after fermentation. See: "*Recept für Schnupftoback* [Recipe for Snuff Tobacco]," 1797, 1 p., and "*Recept für Schnupf-Tobac zu bereiten* [Recipe for Preparing Snuff Tobacco]," no date, 2 pp. (MAS, file box R-705, folder 3); "*Gründliche Anweisung zur ächten und vollkommensten Verfertigung und Zubereitung aller vorzüglichen Rauch- und Schnupftabacke* [Thorough Instructions for the True and Complete Manufacture and Preparation of All the Choice Smoking and Snuff Tobaccos]," Berlin, 1790, pp. 168-171, plate 8 (NYPL); Weiss H. B., and G. M. Weiss, "The Early Snuff Mills of New Jersey," Trenton, 1962, pp. 26-37; Prentis, W. H., "The Snuff-Mill Story. Local History of Morden, Mitcham, Merton," Mitcham, Surrey, 1972, pp. 135-141 (NYPL).
38. ["Single Brethren's Ledger"], G, f. 73, Mar. 15, 1790; f. 74, Mar. 10, 1791 (MAB).
39. "Single Brethren's Diary," Oct. 18, 1779 (MAB). "Our oat dryer caught fire through carelessness in drying tobacco."
40. From 1787 to 1812 the Moravian tobacco manufactory in Bethlehem made regular payments to the oil mill for milling tobacco (note 35). Some of the ledger entries also mention "wood for drying" and the weight of the tobacco processed.
41. From 1787 to 1794 the annual inventories for the Bethlehem tobacco manufactory show stamped tobacco [*gestossenen Tabac*] on hand. From 1795 to 1813 stamped tobacco is absent, but milled tobacco [*gemahlen Toback*] and tobacco dust [*Tobackstaub*] are listed instead. From 1766 to 1793 all payments to the oil mill for processing tobacco refer to stamped or pounded tobacco [*Tabac gestampft* or *Tabac zu stossen*]. Starting in 1794, however, the ledger entries always speak of milled tobacco [*Tobacksblätter gemahlen*] (note 35).
42. "*Inventarium der Oelmühle*," May 31, 1794; May 31, 1795; May 31, 1796. The cost of the new tobacco mill and bolter/sieve was amortized over three years. It seems like more than a coincidence that the tanbark mill needed major repairs at the same time that the new snuff mill was installed.
43. "*Gründliche Anweisung* . . . ," plate 8.
44. Gray, A., "The Experienced Millwright," Edinburgh, 1806, p. 42, plate 17 (EMHL).
45. "Oil Mill Account Book," Aug. 31, 1766. Note that the tanbark stampers were enclosed in a wooden compartment just before tobacco stamping started in the 1765 oil mill ("Oil Mill Diary," Sep. 29-Oct. 11, 1766).
46. Ogden, J. C., "An Excursion into Bethlehem & Nazareth in Pennsylvania in the Year 1799; with a Succinct History of the Society of United Brethren, Commonly Called Moravians," Philadelphia, 1805, p. 21. Gilpin, J., "Journey to Bethlehem" in "Pennsylvania Magazine of History and Biography," vol. 46, Philadelphia, 1922, p. 146.
47. For a short period in the 1790's, snuff mills were taxed by the federal government at a rate dependent upon the type of manufacturing machinery employed. In 1795-1796 taxes were collected from thirteen snuff manufacturing installations in Pennsylvania; one of these was the snuff mill at the Bethlehem oil mill. Eleven of those taxed were "mortars worked by water," one was a "pestle not worked by hand," and one was "worked by stampers and grinders." See: "A Statement of the Revenue arising on Mills and Machinery used in the manufacture of Snuff, from the 1st day of October, 1795, to the 30th September, 1796" in "American State Papers. Documents Legislative and Executive of the Congress of the United States . . . 1789-1815," vol. V ("Finance," vol. I), 1832, p. 564; "*Protokol des Aufseher Collegii*," Sep. 10, Oct. 8, 1794.
48. This drawing of a powered snuff mull is based upon the machinery displayed at the Gilbert Stuart Birthplace, Saunderstown RI. According to Curator A. E. Cummings, this snuff mull was built in 1757, used in Sheffield, England, and brought to the Stuart Birthplace in 1932.
49. Weiss & Weiss, pp. 40-41, 78-79.
50. Only one ledger entry specifically mentions sieving the stamped tobacco at the oil mill (["Single Brethren's Ledger"], G, f. 74, Mar. 10, 1791), but we assume that sieving was a regular practice too obvious to need recording. The installation of a new bolter/sieve in 1794 specifically for use with the new tobacco mill indicates that snuff was sifted after that date (note 42).
51. "Mr. Charles D. Bishop To Rush & Muhlenberg Dr.," Sep. 8, 1832, 1 p. (MAB, file box "Civil Treasury [*Bürgerliche Casse*]. Bills etc.: 1832/33," item 49). For more details of the 1832 waterworks installation at the oil mill, see ["Civil Treasury Journal 1829-1834"], pp. 52-55, 58-59 (MAB, file box "Civil Treasury [*Bürgerliche Casse*]. Journal: 1829-1834. Day Book: 1825-1846").
52. Martin, J. H., "Historical Sketch of Bethlehem in Pennsylvania," Philadelphia, 1872, p. 28. Rau, R., "Historical Sketch of the Bethlehem Waterworks," Bethlehem, 1877, pp. 8-9. International Correspondence Schools, "The Steam Engine, the Indicator, Engine Testing, . . . Pumps," Scranton, 1904, sec. 34, p. 37.
53. Rau, pp. 8-9. "On Friday afternoon . . . " in "The Moravian," Bethlehem, Oct. 30, 1889, p. 700.
54. Worthington Corporation, "Worthington P.S.I. - Pump Selector for Industry," New York, [c. 1970], p. 17-1. Calculations by Stephen Young.
55. "Water! Water!," BDT, Jun. 20, 1867, p. 3. "No Water," BDT, Aug. 19, 1867, p. 3. "No Water," BDT, Mar. 18, 1868, p. 3.
56. ["Minutes of the Sessions of the Board of Managers of the Bethlehem Water Co."], session 10, Aug. 22, 1845; session 11, Sep. 2, 1845; session 14, Feb. 13, 1846 (MAB, file box "Bethlehem Water Co., 1837-1850," folder "Agreements 1838-1850"). "Statement of the books, papers and effects delivered by J. C. Brickenstein former Secretary and Treasurer of the Bethlehem Water Company to Chas. W. Rauch his successor in office. April 13 & 16, 1850," 4 pp., item 29 (MAB, file box "Bethlehem Water Co. 1837-1850," folder "Treasurer 1837-1850"). Rauch, R. S., "Diary of Reuben S. Rauch," 129 pp. (MAB, file box "Bethlehem: Diary of R. S. Rauch 1854-1857"). The Nov. 7, 1856 entry reads: "The new waterwheel at the oil mill is finished and was put into operation yesterday afternoon and seems to work well. Power has been gained, as the pump makes the longer stroke now, whereas the horizontal wheel [turbine] never had power for even the shortest stroke, since a stroke extra short (*sic*) was made for the pump ten years ago, and even for that power was insufficient until a fly wheel was attached." Birkinbine, H. P. M., "Report on the

Water Supply of Bethlehem, Pa.," 1874, p. [1] (MAB, file box "Bethlehem Water Supply ca. 1870"); "Had the same character of pumps . . . been connected with the turbine wheel put in in 1846, it would not have been a failure." Hunter, L. C., "A History of Industrial Power in the United States, 1780-1930. Vol. 1: Waterpower in the Century of the Steam Engine," Charlottesville VA, 1979, pp. 313-318 (description of "Howd's Improved Water Wheel" [turbine]).

57. During the nineteenth century the size of a steam boiler was measured by horsepower. One "rated boiler horsepower" was defined as equivalent to 10 sq. ft. (0.93 sq. m) of boiler heating surface. See Babcock & Wilcox Co., "Steam, Its Generation and Use," 37th ed., New York, 1963, pp. 11-9, 11-10.
58. This location for the boilers is inferred from the location of the chimney in fig. 22.
59. "Trial of the New Pumping Engines," BDT, Jun. 8, 1868, p. 3. "Stated Meeting of Bethlehem Water Board," BDT, Aug. 3, 1875, p. 3.
60. International Correspondence Schools, sec. 34, pp. 5-7. A "noncondensing" steam pump exhausts live steam rather than water condensate.
61. "Special Session of the Bethlehem Water Board," BDT, Apr. 15, 1874, p. 4. "At the Water Works," BDT, Jun. 10, 1874, p. 3.
62. Sanborn-Perris Map Co., "Insurance Maps of Bethlehem, Pennsylvania," New York, 1897, f. 1 (MAB). We believe that the "22 × 36 × 36" dimensions (steam cylinder diameter, water cylinder diameter, stroke length) for the Cameron pump should read "22 × 12 × 36"; see Rau, p. 10. A higher pumping rate is given in a newspaper story ("The Steam Pump," BDT, Jul. 3, 1874, p. 3), but the rate quoted by Rau is probably more reliable.
63. International Correspondence Schools, sec. 35, pp. 13-15. Since the air in the chamber was gradually absorbed by the water, some means of renewal was required. Hence air compressors of some form were required auxiliary equipment for all direct-acting pumps ("Town Council of Bethlehem Borough," SBG, Mar. 12, 1912, p. 1).
64. Sanborn-Perris Map Co., "Bethlehem, Pennsylvania," 1892, f. 1 (MAB). There is a discrepancy among various sources concerning the capacity of this pump. Critics of the installation (BDT, Jun. 11, 1889, p. 1) later claimed the capacity to be only 350 gallons per minute (gpm). A list of installations in the Worthington catalog (note 65, p. 29) rates the Bethlehem pump at a nominal 520 gpm. But in the same catalog the Worthington 12 × 8.5 × 10 duplex pump, the size installed at Bethlehem, is rated at 245 to 490 gpm (p. 6). These differences may be rectified by considering the 520 gpm figure as a nominal rating for comparison with other installations. The range given in the catalog listing reflects real operating conditions, where the output is dependent upon the pressure and volume of the steam supplied to the pump. The measured flow of 350 gpm indicates that the steam supply was somewhat low for the head (resistance to flow in the water piping system) against which the pump had to work. Thus both the manufacturer's figures and the measured flow were correct, but under different conditions.
65. Henry R. Worthington [Co.], "Worthington Steam Pumping Machinery," New York, 1887, pp. 4-5 (EMHL).
66. "Bethlehem Town Council," BDT, Dec. 27, 1888, p. 1. "The Water Works Mishap," BDT, Mar. 13, 1889, p. 1. "A special meeting . . . " in "The Moravian," Bethlehem, Jan. 2, 1889, p. 12.
67. "Local News Notes. The new stack . . . ," BDT, Mar. 23, 1889, p. 1. This article specifies a 60 foot chimney, while the Sanborn insurance maps (fig. 23) estimated it to be 70 (1892) or 90 (1897) feet high.
68. This small direct-acting steam pump had a 4.5 inch (11.4 cm) diameter steam piston, a 3 inch (7.6 cm) water piston, and a 4 inch (10.2 cm) stroke ("Bethlehem Town Council," BDT, Dec. 27, 1888, p. 1). Although described as "a pump . . . to feed the pumps," it most likely was the usual boiler water feed pump.
69. "North Side Town Council, The Water Works," BDT, May 7, 1889, pp. 1,4. "Bethlehem's Water Works," BDT, Oct. 26, 1889, p. 1. "Bethlehem Water Works. The New and Complete Plant Now in Use," BDT, Jan. 25, 1890, p. 4.
70. Deane Steam Pump Co., "The Deane Steam and Power Pumps and Pumping Machinery," Holyoke MA, 1897, pp. 70-71 (EMHL).
71. International Correspondence Schools, sec. 34. pp. 20-21.
72. "Memorablia of 1911 at Year's Final Service," SBG, Jan. 1, 1912, p. 6.
73. "Bethlehem Water Works," SBG, Dec. 7, 1911, p. 1. "Water Supply Permit for Borough of Bethlehem," SBG, Mar. 8, 1912, p. 1. "Bethlehem Borough Council Session," SBG, Jan. 9, 1912, p. 1.
74. "New Centrifugal Pump," SBG, Nov. 7, 1912, p. 1. "Used Old Works Tank," SBG, Jan. 3, 1913, p. 2. "Bethlehem Borough Water," SBG, Jan 10, 1913, p. 1.
75. "Bethlehem Water Works," SBG, Dec. 7, 1911, p. 1. The "De Lavalle" centrifugal pump mentioned here was undoubtedly a De Laval pump; see De Laval Steam Turbine Co., "De Laval High Efficiency Centrifugal Pumps," Trenton, 1910, 96 pp. (EMHL). "Bethlehem Borough Council Session," SBG, Nov. 12, 1912, p. 1.
76. "Bethlehem Borough Council Session," SBG, Sep. 10, 1912, p. 1. "New Centrifugal Pump," SBG, Nov. 7, 1912, p. 1.
77. Acrelius, I., "A History of New Sweden" in "Memoirs of the Historical Society of Pennsylvania," vol. 11, Philadelphia, 1874, p. 400. [Kriebel, H. W.], "Perkiomen Valley Water Power" in "Perkiomen Region," vol. 8, no. 1, Pennsburg PA, 1930, pp. 10-13 (Swenkfelder Library, Pennsburg PA). Berky, A. S., "An Account of Some Hosensack Valley Mills," Pennsburg PA, 1957, 70 pp. (EMHL).
78. Smeaton, J., "Reports of the Late John Smeaton," vol. 2, London, 1812, plate 16. Wailes, R., "The English Windmill," London, 1954, plate 9. Tann, J., "The Development of the Factory," London, 1970, pp. 56-58.
79. Quincy, E., "A Treatise of Hemp-Husbandry," Boston, 1765, pp. 31-32, plate (EMHL). Wissett, R., "A Treatise on Hemp," London, 1808, p. 268.
80. Warner, M. E., "The Old Tanbark Mill" in "Chronicle of the Early American Industries Association," vol. 1, New York, p. 18-6. Landis, H. K., "Tan-Bark Mills," ibid., pp. 24-7, 24-8. Welsh, P. C., "Tanning in the United States to 1850," Washington, 1964, pp. 2, 8.
81. Prentis, pp. 135-141. Tann, p. 96.

CONSTRUCTION RECORDS FOR THE 1765 OIL MILL

1. "Oil Mill Diary," Dec. 10-12, 1764. "Oil Mill Account Book," Mar. 31, 1766. [Arbo, J., "Cost of Material and Work in Various Building Operations"], 1761-1770, p. [51] (MAB, file box "Bethlehem: *Bau-Inspection.* Building Material: 1761-1766 . . . ").
2. "*Conferenz Protocoll des Gemein-Raths in Bethlehem 1762* [Minutes of the Congregation Council in Bethlehem 1762]," p. [28] (MAB).
3. "Oil Mill Diary," Dec. 17, 1764.
4. We have examined "*Haupt-Buch vor Christoph Christensen, Mühl-Bauer in Bethlehem* [Ledger for Chris-

toph Christensen, Mill Builder in Bethlehem]" (MAB, file box "TR. Mill & Wheelwright") and find the handwriting very similar to that of John Arbo.

5. [Arbo, J., "Revised Oil Mill Account Book"], financial journal starting Dec. 31, 1764, p. [10] (MAB, file box "*Oeconomische Tage-Bücher*: 1756-1762, 1764-1776, . . . ").
6. "*Haupt-Buch vor die Bau-Inspection in Bethlehem 1762-1766* [Ledger for Building Supervision in Bethlehem 1762-1766]," f. 140 (MAB, file box "Bethlehem: *Bau-Inspection. Haupt-Buch*: 1762-1766 & 1766-1803").
7. "Oil Mill Diary," Jul. 8, 16, 1766. "*Chor-Diarium der Ledigen Brüder* [Single Brethren's Choir Diary]," Jul. 8, 16, 1766 (MAB).
8. The short biography of John Arbo on pp. 26-28 refers to some of his known manuscripts.
9. Arbo, J., ["John Arbo's Commonplace Book 1761-1765"], pp. [14, 44] (MAB).
10. [Arbo, J.], "*Diarium den Bau der Oel-Mühle betr.* [Construction Diary for the Oil Mill]," elsewhere called the "Oil Mill Diary," 1764-1772, 110 pp. (MAB, file box "Oil Mill. Building of the Oil Mill 1764-1784").
11. "The Oil Mill Diary," translation by Hans-Joachim Finke, [1976] (MAB). "Translation. Diary re Building of the Oil Mill 1764-1772," translation by Robert Rau, 2 vols., [c. 1880] (MAB, Robert Rau Collection, file box "Municipal Affairs, Industry, Trades, etc.").
12. To our knowledge this plan has not survived.
13. Text inside parentheses appears as a footnote in the original manuscript.
14. A 1761 map of land ownership in the Bethlehem area shows the Ostrom (Astrom) property located c. 1 mile (1.5 km) southwest of the oil mill site on the south side of the Lehigh River. See Yates et al., "Bethlehem of Pennsylvania: The First One Hundred Years," Bethlehem, 1968, p. 99.
15. The large horse-drawn sleds of stones hauled to the oil mill are counted here, not the small handsleds used to move the stones across the Lehigh. A fourth column for tallying stones belonging to the Single Brethren's House has no entries.
16. The source of these stones is described in the "Oil Mill Account Book," Feb. 7, 1765.
17. Probably the Irish Settlement 6 miles (10 km) northwest of Bethlehem, now the village of Weaversville.
18. *Kalkhütte* can be translated as either "lime kiln" or "lime shed." The latter meaning is more likely since the structure was put together with nails ("Oil Mill Account Book," Jun. 11, 1765).
19. This new channel was located slightly west of the old one. The map in fig. 56 shows the 1765 oil mill located right on the eastern bank of Monocacy Creek.
20. For the new waterwheel ("Oil Mill Diary," May 7, 1765).
21. An earthen dam between the old and new channels of Monocacy Creek ("Oil Mill Diary," Jul. 29, 1765).
22. The foundation of the burned 1752 oil mill. The 1765 oil mill was built on the same site.
23. From the burned 1752 oil mill.
24. Bailing and pumping were required to remove the water seeping into the foundation trenches.
25. The Moravian settlement at what is now Winston-Salem, North Carolina.
26. Nazareth and its outlying farming communities.
27. The exact nature of this warranty (*mit Condition*) is unknown. Perhaps it guaranteed a certain yield of linseed oil or part of the purchase price would be refunded.
28. The old dam for the 1752 oil mill was located in the old abandoned course for Monocacy Creek. The new dam constructed for the 1765 oil mill was located in the newly-dug course.
29. *Dürchzuge*, probably the horizontal, floor-supporting crossbeams. See Nov. 20, 1765 entry.
30. Probably his birthday.
31. The dam at the front of the new head race. This probably was part of the earthen dam erected to separate the old and new channels of Monocacy Creek.
32. Anonymous, "Waterworks Dam & Creek 1870," photograph, 1870 (HBI P-62-2-779A).
33. *Haus*, presumably the small entrance wing containing the *Vor-Haus* [entry] and the *Stube* [room] (see fig. 29).
34. By German convention the floor levels of a building are counted: ground floor, first floor, second floor, etc. Hence the "second floor level" was at the eaves of the oil mill roof.
35. Presumably a set of posts and crosspieces that served as a workstand for assembling the various parts of the waterwheel. See Reynolds, T. S., "Stronger Than a Hundred Men: A History of the Vertical Water Wheel," Baltimore, 1983, p. 164.
36. Text lists Monday, Tuesday, and Wednesday as Dec. 1, 2, & 3, which is incorrect. Since the Moravians avoided Sunday work where possible, we presume that Dec. 2, 3, & 4 were the intended dates.
37. Although "rubbing hemp" is the English term used by the diarist, the hemp mill machinery in 1765 could only have been used for stamping hemp (see figs. 29 & 30).
38. Both English and German versions of this advertisement are given by the diarist. The English version is quoted verbatim here.
39. More often spelled Buz or Butz.
40. On the bottoms of the stamping mortars in the oilseed mill.
41. The pan for roasting crushed flaxseed in the oilseed mill.
42. Probably the *Quetsch-Mühle* [crushing mill] next to the wedge press in the oilseed mill (fig. 29).
43. This sheet metal may have been used in the clapper mill, which could be thought of as part of the groat mill. Note that the clapper mill was completed the following week.
44. Text labels this day Monday, but Apr. 1, 1766 was a Tuesday.
45. Dr. John Matthew Otto or Dr. John Frederick Otto, local physicians and apothecaries.
46. A calculation of Christensen's weekly wages. See "Oil Mill Diary," Dec. 12, 1764.
47. This water box and the water trough mentioned below were for the town water system and were not part of the oil mill project.
48. The two combination face/lantern gears in the groat mill drive (fig. 28).
49. The Single Brethren's Choir held an annual festival on August 29.
50. The amounts owed by Rauschenberger and Otto are crossed out, no doubt indicating that the debts were paid.
51. Probably a bell or clapper that sounded automatically when the grain hopper was empty.
52. This "dryer" may have been a device or a flat area in the actual groat mill for drying grain and tobacco prior to milling. The separate oat drying kiln building was not built until October 1767 ("Oil Mill Account Book," Oct. 10, 1767).
53. Should be 10½ days.
54. *Vulcanus* is probably an oblique reference to an unpleasant incident in the Single Brethren's Choir ("Single Brethren's Choir Diary," Jan. 2, 1767).
55. The diarist has "13" here, but Thursday was clearly the 15th.
56. Leupold, J., and J. M. Beyern, "*Theatrum machinarum*

molarium, oder Schau-Platz der Mühlen-Bau-Kunst [A Theater of Mill Machinery]," 2nd ed., Dresden, 1767, part I, plate XVIII, fig. 5 (double wing gudgeon). Evans, O., "The Young Mill-Wright and Miller's Guide," 5th ed., Philadelphia, 1826, plate XI, fig. 6 (four wing gudgeon).

57. Winnowing machine for cleaning grain (fig. 43).
58. Number crossed out in original. There is a discrepancy in the calculation of board expenses for Leimbach and Christensen.
59. Both numbers crossed out in original manuscript.
60. [Arbo, J.], "*Berechnung der Bau Unkosten bei der Oel-Mühle und Brauerei* [Account of Construction Costs for the Oil Mill and Brewery]," elsewhere called the "Oil Mill Account Book," 1764-1768 (oil mill, attributed to Arbo), 1781-1784 (brewery, author unknown), pp. 28-[47, 63-67] (MAB, file box "Oil Mill. Building of the Oil Mill, 1764-1784").
61. "*Oel Mühlbau Conto* [Oil Mill Building Account]," ["Single Brethren's Ledger"], B, ff. 211-213, 243-244 (missing); C, ff. 152-153, 259 (MAB).
62. "Oil Mill Account Book," pp. 28-[47].
63. Salvaged from the burned 1752 oil mill.
64. One perch equals 24.75 ft^3 (0.70 m^3).
65. The inventory of equipment salvaged from the burned 1752 oil mill.
66. Since these [metal] rollers were in the linseed mill (Mar. 11, 1765 entry), were held in a framework, and were associated with a lantern gear (p. 97), they may possibly have been the crushing rollers in the small mill (*Quetsch-Mühle*) for grinding presscake. Fig. 29 shows a lantern gear in the drive train of this mill.
67. The waterwheel salvaged from the burned 1752 oil mill.
68. This figure is marked with an asterisk in the original text to relate it to the Feb. 28, 1766 entry.
69. *Vereniging de Zaansche Molen*, "*De Geschiedenis van De Bonte Hen* [The History of *De Bonte Hen* Mill]," Koog a/d Zaan, Netherlands, 1975, p. 13.
70. Bags to hold the crushed, roasted flaxseed in the wedge press at the oilseed mill (see fig. 35).
71. The pan for roasting crushed flaxseed in the oilseed mill (see fig. 34).
72. This hair from the tanner Geitner was probably used for caulking the raceway.
73. Our translation of *Mühlpfanne* and *Pfanne* as "bearing" is based upon the context of the usage here. It was a brass fitting placed underneath the end of the gudgeon rod.
74. Refers to Aug. 31, 1766 entry.
75. The trough in the wedge press at the oilseed mill (see fig. 35).
76. ["Single Brethren's Ledger"], C, f. 270.
77. Perforated iron plates covering the floor of the drying kiln.
78. The hair cloth covering the drying floor probably was held in place with hooks and rings.
79. *Spindeln* refers to rounded pieces of wood. Perhaps these were rungs for a lantern gear.
80. Iron bands.
81. Litchfield, C., "Hemp Roller Stone (no. 11149) at Mercer Museum, Doylestown PA," photograph, 1981.
82. "Oil Mill Account Book," pp. [63-67]. An earlier calculation of this cost summary appears in one of Arbo's notebooks. See [Arbo, J., "Cost of Material and Work in Various Building Operations"], 1761-1770, pp. [21-25] (MAB, file box "Bethlehem: *Bau-Inspection*. Building Material: 1761-1766 . . . ").
83. An earlier "Recapitulation" dated Oct. 21, 1766 appears on the same page, but it totals £1073.11.11 indicating that some costs were not included.
84. "Oil Mill Diary," Oct. 18, 1765.
85. [Arbo, J., "Cost of Material and Work in Various Building Operations"], pp. [50-52].
86. Ibid., p. [50]. Figure traced with ink for clearer reproduction.
87. Arbo's calculations define *Ruthen* as 24.75 ft^3 (0.70 m^3), the standard volume for one British perch of stones. This is equivalent to a pile of stones one rod (16½' or 5.03 m) long, 1' high, and 1½' thick. Values in the "Perches" column are expressed in 99ths because Arbo multiplied both the ft^3 value and its 24.75 divisor by 4 before performing long division.
88. I.e., the footing.
89. By German convention the "first floor" refers to the floor above ground level.
90. The original manuscript has "EF" here, but clearly "CD" is the 2½' high wall referred to (see figs. 30 & 31).

APPENDIX. CHRISTENSEN AND ARBO MEMOIRS

1. "*Chor-Diarium der Ledigen Brüder* [Single Brethren's Choir Diary]," Sep. 18, 1776 (MAB).
2. Matthias Christensen, 1707-1787 (Christiansfeld Moravian Archives, memoirs P.A.II-R7.10 and P.A.II-R7.18).
3. Near Winston-Salem, North Carolina.
4. "*Personalia des seligen Br. Joh. Arboe* [Personal Account of the Late Brother John Arbo]," 9 pp. (MAB, file box "Biographies of Moravian Clergymen and Workers. Box I," folder I).
5. Zinzendorf, L. von, "*Inhalt dererjenigen Reden Welche zu Berlin von 1ten Januarii 1738 biß 27ten Aprilis in denen Abend-Stunden sonderlich für die Manns-Personen gehalten worden* [Contents of Those Sermons Delivered Especially for the Men in Berlin from January 1 to April 27 in the Afternoon]," n.p., 1738, 342 pp. (MAB).
6. A continuous prayer vigil.

INDEX

Page numbers followed by "f" or "n" refer to a figure or to a note. Thus "33f" refers to a figure on page 33, and "110n11" refers to note 11 on page 110.

ACKNOWLEDGMENTS

We are greatly indebted to Rev. Vernon H. Nelson and Dr. Lothar Madeheim at the Moravian Archives in Bethlehem for their many helpful consultations and discussions throughout our research studies. Without their extensive knowledge of the manuscript collection at the Archives, many documents relating to the Bethlehem oil mill would never have been found.

We thank mill historian Stephen Kindig for his assistance in interpreting the hemp and groat milling operations in the 1765 oil mill. His continued interest in the progress of our research has helped us to maintain our enthusiasm for this book project over a period of seven years.

Special appreciation goes to Harry Rinker, who first told us about the extensive early Moravian records on the Bethlehem oil mill. His letter of August 3, 1973 started the chain of events that eventually lead to the publication of this book.

We also wish to thank Robert T. Brown, Susan Finkel, Grethe Goodwin, Charles Howell, Dr. Carol Litchfield, Dr. Leslie O'Malley, Dr. Bruce Seely, Dr. Theodorus van Es, and Joan S. Young for their consultation and advice on various details of our manuscript.

We are grateful to the following organizations for permission to photograph items in their collections for the figures in this book:

Archiv der Brüder Unität, figures 28-31.
Bucks County Historical Society, figure 63.
Eleutherian Mills Historical Library, figures 50-52, 60.
Freilichtmuseum Vogtsbauernhof, figure 39.
Historic Bethlehem Inc., figures 1, 5-8, 24, 27, 57.
Library Company of Philadelphia, figure 15.
Moravian Archives (Bethlehem), figures 2-4, 9-12, 14, 17-22, 25-26, 53-56, 58-59, 61, 64.
New York Public Library, Astor, Lenox, and Tilden Foundations, Arents Collections, figure 48.
Princeton University Library, figure 13.
Vereniging de Zaansche Molen, figure 62.

COLOPHON

CARTER LITCHFIELD is a specialist in the chemical and technological history of fatty materials and a former Professor of Biochemistry at Rutgers University in New Brunswick, New Jersey.

HANS-JOACHIM FINKE is Program Director of the Regional Conference of Historical Agencies at Manlius, New York. His research, writing, and teaching activities have focused on the history of Germany and the early German immigrants to Pennsylvania.

STEPHEN G. YOUNG is a mechanical engineer with the Bethlehem Steel Corporation. He served as technical consultant in the restoration of the 1762 waterworks and the 1869 grist mill at Bethlehem and is a member of the Board of Directors of Historic Bethlehem Inc.

KAREN ZERBE HUETTER is a former Director of Museum Services for Historic Bethlehem Inc. and the author of a booklet on the history of the Bethlehem waterworks.

This book was edited and designed by Olearius Editions, which publishes books on historic oil mills. Its pressmark, which appears above, is the cross-and-circle symbol used by medieval alchemists to designate olive oil, one of the most ancient of fatty oils used by man.

Fifteen hundred copies of this edition were printed and bound by the J. D. Lucas Printing Co. at Baltimore MD.